MADAME HOW AND LADY WHY

OR, FIRST LESSONS IN EARTH LORE FOR CHILDREN

By CHARLES KINGSLEY

Madame How and Lady Why
By Charles Kingsley

Print ISBN 13: 978-1-4209-7020-3
eBook ISBN 13: 978-1-4209-7021-0

This edition copyright © 2020. Digireads.com Publishing.

Cover Image: a detail of "Vesuvius Erupting", by Pierre Jacques Volaire (1729-c. 1802) / Bridgeman Images.

Please visit *www.digireads.com*

CONTENTS

DEDICATION

TO

MY SON GRENVILLE ARTHUR,

AND TO

HIS SCHOOL-FELLOWS AT WINTON HOUSE

THIS LITTLE BOOK IS DEDICATED.

Preface

My dear boys,—When I was your age, there were no such children's books as there are now. Those which we had were few and dull, and the pictures in them ugly and mean: while you have your choice of books without number, clear, amusing, and pretty, as well as really instructive, on subjects which were only talked of fifty years ago by a few learned men, and very little understood even by them. So if mere reading of books would make wise men, you ought to grow up much wiser than us old fellows. But mere reading of wise books will not make you wise men: you must use for yourselves the tools with which books are made wise; and that is—your eyes, and ears, and common sense.

Now, among those very stupid old-fashioned boys' books was one which taught me that; and therefore I am more grateful to it than if it had been as full of wonderful pictures as all the natural history books you ever saw. Its name was *Evenings at Home*; and in it was a story called "Eyes and no Eyes;" a regular old-fashioned, prim, sententious story; and it began thus:—

"Well, Robert, where have you been walking this afternoon?" said Mr. Andrews to one of his pupils at the close of a holiday.

Oh—Robert had been to Broom Heath, and round by Camp Mount, and home through the meadows. But it was very dull. He hardly saw a single person. He had much rather have gone by the turnpike-road.

Presently in comes Master William, the other pupil, dressed, I suppose, as wretched boys used to be dressed forty years ago, in a frill collar, and skeleton monkey-jacket, and tight trousers buttoned over it, and hardly coming down to his ankles; and low shoes, which always came off in sticky ground; and terribly dirty and wet he is: but he never (he says) had such a pleasant walk in his life; and he has brought home his handkerchief (for boys had no pockets in those days much bigger than key-holes) full of curiosities.

He has got a piece of mistletoe, wants to know what it is; and he has seen a woodpecker, and a wheat-ear, and gathered strange flowers on the heath; and hunted a peewit because he thought its wing was broken, till of course it led him into a bog, and very wet he got. But he did not mind it, because he fell in with an old man cutting turf, who told him all about turf-cutting, and gave him a dead adder. And then he went up a hill, and saw a grand prospect; and wanted to go again, and make out the geography of the country from Cary's old county maps, which were the only maps in those days. And then, because the hill was called Camp Mount, he looked for a Roman camp, and found one; and then he went down to the river, saw twenty things more; and so on, and so on, till he had brought home curiosities enough, and thoughts enough, to last him a week.

Whereon Mr. Andrews, who seems to have been a very sensible old gentleman, tells him all about his curiosities: and then it comes out—if you will believe it—that Master William has been over the very same ground as Master Robert, who saw nothing at all.

Whereon Mr. Andrews says, wisely enough, in his solemn old-fashioned way,—

"So it is. One man walks through the world with his eyes open, another with his eyes shut; and upon this difference depends all the superiority of knowledge which one man acquires over another. I have known sailors who had been in all the quarters of the world, and could tell you nothing but the signs of the tippling-houses, and the price and quality of the liquor. On the other hand, Franklin could not cross the Channel without making observations useful to mankind. While many a vacant thoughtless youth is whirled through Europe without gaining a single idea worth crossing the street for, the observing eye and inquiring mind find matter of improvement and delight in every ramble. You, then, William, continue to use your eyes. And you, Robert, learn that eyes were given to you to use."

So said Mr. Andrews: and so I say, dear boys—and so says he who has the charge of you—to you. Therefore I beg all good boys among you to think over this story, and settle in their own minds whether they will be eyes or no eyes; whether they will, as they grow up, look and see for themselves what happens: or whether they will let other people look for them, or pretend to look; and dupe them, and lead them about—the blind leading the blind, till both fall into the ditch.

I say "good boys;" not merely clever boys, or prudent boys: because using your eyes, or not using them, is a question of doing Right or doing Wrong. God has given you eyes; it is your duty to God to use them. If your parents tried to teach you your lessons in the most agreeable way, by beautiful picture-books, would it not be ungracious, ungrateful, and altogether naughty and wrong, to shut your eyes to those pictures, and refuse to learn? And is it not altogether naughty and wrong to refuse to learn from your Father in Heaven, the Great God who made all things, when he offers to teach you all day long by the most beautiful and most wonderful of all picture-books, which is simply all things which you can see, hear, and touch, from the sun and stars above your head to the mosses and insects at your feet? It is your duty to learn His lessons: and it is your interest. God's Book, which is the Universe, and the reading of God's Book, which is Science, can do you nothing but good, and teach you nothing but truth and wisdom. God did not put this wondrous world about your young souls to tempt or to mislead them. If you ask Him for a fish, he will not give you a serpent. If you ask Him for bread, He will not give you a stone.

So use your eyes and your intellect, your senses and your brains, and learn what God is trying to teach you continually by them. I do not mean that you must stop there, and learn nothing more. Anything but that. There are things which neither your senses nor your brains can tell you; and they are not only more glorious, but actually more true and more real than any things which you can see or touch. But you must begin at the beginning in order to end at the end, and sow the seed if you wish to gather the fruit. God has ordained that you, and every child which comes into the world, should begin by learning something of the world about him by his senses and his brain; and the better you learn what they can teach you, the more fit you will be to learn what they cannot teach you. The more you try now to understand *things*, the more you will be able hereafter to understand men, and That which is above men. You began to find out that truly Divine mystery, that you had a mother on earth, simply by lying soft and warm upon her bosom; and so (as Our Lord told the Jews of old) it is by watching the common natural things around you, and considering the lilies of the field, how they grow, that you will begin at least to learn that far Diviner mystery, that you have a Father in Heaven. And so you will be delivered (if you will) out of the tyranny of darkness, and distrust, and fear, into God's free kingdom of light, and faith, and love; and will be safe from the venom of that tree which is more deadly than the fabled upas of the East. Who planted that tree I know not, it was planted so long ago: but surely it is none of God's planting, neither of the Son of God: yet it grows in all lands and in all climes, and sends its hidden suckers far and wide, even (unless we be watchful) into your hearts and mine. And its name is the Tree of Unreason, whose roots are conceit and ignorance, and its juices folly and death. It drops its venom into the finest brains; and makes them call sense, nonsense; and nonsense, sense; fact, fiction; and fiction, fact. It drops its venom into the tenderest hearts, alas! and makes them call wrong, right; and right, wrong; love, cruelty; and cruelty, love. Some say that the axe is laid to the root of it just now, and that it is already tottering to its fall: while others say that it is growing stronger than ever, and ready to spread its upas-shade over the whole earth. For my part, I know not, save that all shall be as God wills. The tree has been cut down already again and again; and yet has always thrown out fresh shoots and dropped fresh poison from its boughs. But this at least I know: that any little child, who will use the faculties God has given him, may find an antidote to all its poison in the meanest herb beneath his feet.

There, you do not understand me, my boys; and the best prayer I can offer for you is, perhaps, that you should never need to understand me: but if that sore need should come, and that poison should begin to spread its mist over your brains and hearts, then you will be proof against it; just in proportion as you have used the eyes and the common sense which God has given you, and have considered the lilies of the field, how they grow.

C. KINGSLEY.

Chapter I. The Glen

You find it dull walking up here upon Hartford Bridge Flat this sad November day? Well, I do not deny that the moor looks somewhat dreary, though dull it need never be. Though the fog is clinging to the fir-trees, and creeping among the heather, till you cannot see as far as Minley Corner, hardly as far as Bramshill woods—and all the Berkshire hills are as invisible as if it was a dark midnight—yet there is plenty to be seen here at our very feet. Though there is nothing left for you to pick, and all the flowers are dead and brown, except here and there a poor half-withered scrap of bottle-heath, and nothing left for you to catch either, for the butterflies and insects are all dead too, except one poor old Daddy-long-legs, who sits upon that piece of turf, boring a hole with her tail to lay her eggs in, before the frost catches her and ends her like the rest: though all things, I say, seem dead, yet there is plenty of life around you, at your feet, I may almost say in the very stones on which you tread. And though the place itself be dreary enough, a sheet of flat heather and a little glen in it, with banks of dead fern, and a brown bog between them, and a few fir-trees struggling up—yet, if you only have eyes to see it, that little bit of glen is beautiful and wonderful,—so beautiful and so wonderful and so cunningly devised, that it took thousands of years to make it; and it is not, I believe, half finished yet.

How do I know all that? Because a fairy told it me; a fairy who lives up here upon the moor, and indeed in most places else, if people have but eyes to see her. What is her name? I cannot tell. The best name that I can give her (and I think it must be something like her real name, because she will always answer if you call her by it patiently and reverently) is Madam How. She will come in good time, if she is called, even by a little child. And she will let us see her at her work, and, what is more, teach us to copy her. But there is another fairy here likewise, whom we can hardly hope to see. Very thankful should we be if she lifted even the smallest corner of her veil, and showed us but for a moment if it were but her finger tip—so beautiful is she, and yet so awful too. But that sight, I believe, would not make us proud, as if we had had some great privilege. No, my dear child: it would make us feel smaller, and meaner, and more stupid and more ignorant than we had ever felt in our lives before; at the same time it would make us wiser than ever we were in our lives before—that one glimpse of the great glory of her whom we call Lady Why.

But I will say more of her presently. We must talk first with Madam How, and perhaps she may help us hereafter to see Lady Why. For she is the servant, and Lady Why is the mistress; though she has a Master over her again—whose name I leave for you to guess. You have heard it often already, and you will hear it again, for ever and ever.

But of one thing I must warn you, that you must not confound Madam How and Lady Why. Many people do it, and fall into great mistakes thereby,—mistakes that even a little child, if it would think, need not commit. But really great philosophers sometimes make this mistake about Why and How; and therefore it is no wonder if other people make it too, when they write children's books about the wonders of nature, and call them "Why and Because," or "The Reason Why." The books are very good books, and you should read and study them: but they do not tell you really "Why and Because," but only "How and So." They do not tell you the "Reason Why" things happen, but only "The Way in which they happen." However, I must not blame these good folks, for I have made the same mistake myself often, and may do it again: but all the more shame to me. For see—you know perfectly the difference between How and Why, when you are talking about yourself. If I ask you, "Why did we go out to-day?" You would not answer, "Because we opened the door." That is the answer to "How did we go out?" The answer to Why did we go out is, "Because we chose to take a walk." Now when we talk about other things beside ourselves, we must remember this same difference between How and Why. If I ask you, "Why does fire burn you?" you would answer, I suppose, being a little boy, "Because it is hot;" which is all you know about it. But if you were a great chemist, instead of a little boy, you would be apt to answer me, I am afraid, "Fire burns because the vibratory motion of the molecules of the heated substance communicates itself to the molecules of my skin, and so destroys their tissue;" which is, I dare say, quite true: but it only tells us how fire burns, the way or means by which it burns; it does not tell us the reason why it burns.

But you will ask, "If that is not the reason why fire burns, what is?" My dear child, I do not know. That is Lady Why's business, who is mistress of Mrs. How, and of you and of me; and, as I think, of all things that you ever saw, or can see, or even dream. And what her reason for making fire burn may be I cannot tell. But I believe on excellent grounds that her reason is a very good one. If I dare to guess, I should say that one reason, at least, why fire burns, is that you may take care not to play with it, and so not only scorch your finger, but set your whole bed on fire, and perhaps the house into the bargain, as you might be tempted to do if putting your finger in the fire were as pleasant as putting sugar in your mouth.

My dear child, if I could once get clearly into your head this difference between Why and How, so that you should remember them steadily in after life, I should have done you more good than if I had given you a thousand pounds.

But now that we know that How and Why are two very different matters, and must not be confounded with each other, let us look for Madam How, and see her at work making this little glen; for, as I told you, it is not half made yet. One thing we shall see at once, and see it more and more clearly the older we grow; I mean her wonderful patience and diligence. Madam How is never idle for an instant. Nothing is too great or too small for her; and she keeps her work before her eye in the same moment, and makes every separate bit of it help every other bit. She will keep the sun and stars in order, while she looks after poor old Mrs. Daddy-long-legs there and her eggs. She will spend thousands of years in building up a mountain, and thousands of years in grinding it down again; and then carefully polish every grain of sand which falls from that mountain, and put it in its right place, where it will be wanted thousands of years hence; and she will take just as much trouble about that one grain of sand as she did about the whole mountain. She will settle the exact place where Mrs. Daddy-long-legs shall lay her eggs, at the very same time that she is settling what shall happen hundreds of years hence in a stair millions of miles away. And I really believe that Madam How knows her work so thoroughly, that the grain of sand which sticks now to your shoe, and the weight of Mrs. Daddy-long-legs' eggs at the bottom of her hole, will have an effect upon suns and stars ages after you and I are dead and gone. Most patient indeed is Madam How. She does not mind the least seeing her own work destroyed; she knows that it must be destroyed. There is a spell upon her, and a fate, that everything she makes she must unmake again: and yet, good and wise woman as she is, she never frets, nor tires, nor fudges her work, as we say at school. She takes just as much pains to make an acorn as to make a peach. She takes just as much pains about the acorn which the pig eats, as about the acorn which will grow into a tall oak, and help to build a great ship. She took just as much pains, again, about the acorn which you crushed under your foot just now, and which you fancy will never come to anything. Madam How is wiser than that. She knows that it will come to something. She will find some use for it, as she finds a use for everything. That acorn which you crushed will turn into mould, and that mould will go to feed the roots of some plant, perhaps next year, if it lies where it is; or perhaps it will be washed into the brook, and then into the river, and go down to the sea, and will feed the roots of some plant in some new continent ages and ages hence: and so Madam How will have her own again. You dropped your stick into the river yesterday, and it floated away. You were sorry, because it had cost you a great deal of trouble to cut it, and peel it, and carve a head and your name on it. Madam How was not sorry, though she had taken a great deal more trouble with that stick than ever you had taken. She had been three years making that stick, out of many things, sunbeams among the rest. But when it fell into the river, Madam How knew that she should not lose her sunbeams nor anything else: the stick would float down the river, and on into the sea; and there, when it got heavy with the salt water, it would sink, and lodge, and be buried, and perhaps ages hence turn into coal; and ages after that some one would dig it up and burn it, and then out would come, as bright warm flame, all the sunbeams that were stored away in that stick: and so Madam How would have her own again. And if that should not be the fate of your stick, still something else will happen to it just as useful in the long run; for Madam How never loses anything, but uses up all her scraps and odds and ends somehow, somewhere, somewhen, as is fit and proper for the Housekeeper of the whole Universe. Indeed, Madam How is so patient that some people fancy her stupid, and think that, because she does not fall into a passion every time you steal her sweets, or break her crockery, or disarrange her furniture, therefore she does not care. But I advise you as a little boy, and still more when you grow up to be a man, not to get that fancy into your head; for you will find that, however good-natured and patient Madam How is in most matters, her keeping silence and not seeming to see you is no sign that she has forgotten. On the contrary, she bears a grudge (if one may so say, with all respect to her) longer than any one else does; because she will always have her own again. Indeed, I sometimes think that if it were not for Lady Why, her mistress, she might bear some of her grudges for ever and ever. I have seen men ere now damage some of Madam How's property when they were little boys, and be punished by her all their lives long, even though she had mended the broken pieces, or turned them to some other use. Therefore I say to you, beware of Madam How. She will teach you more kindly, patiently, and tenderly than any mother, if you want to learn her trade. But if, instead of learning her trade, you damage her materials and play with her tools, beware lest she has her own again out of you.

Some people think, again, that Madam How is not only stupid, but ill-tempered and cruel; that she makes earthquakes and storms, and famine and pestilences, in a sort of blind passion, not caring where they go or whom they hurt; quite heedless of who is in the way, if she wants to do anything or go anywhere. Now, that Madam How can be very terrible there can be no doubt: but there is no doubt also that, if people choose to learn, she will teach them to get out of her way whenever she has business to do which is dangerous to them. But as for her being cruel and unjust, those may believe it who like. You, my dear boys and girls, need not believe it, if you will only trust to Lady Why; and be sure that Why is the mistress and How the servant, now and for ever. That Lady Why is utterly good and kind I know full well; and I believe that, in her case too, the old proverb holds, "Like mistress, like servant;" and that the more we know of Madam How, the more we shall be content with her, and ready to submit to whatever she does: but not with that stupid resignation which some folks preach who do not believe in lady Why—that is no resignation at all. That is merely saying—

"What can't be cured
Must be endured,"

like a donkey when he turns his tail to a hail-storm,—but the true resignation, the resignation which is fit for grown people and children alike, the resignation which is the beginning and the end of all wisdom and all religion, is to believe that Lady Why knows best, because she herself is perfectly good; and that as she is mistress over Madam How, so she has a Master over her, whose name—I say again—I leave you to guess.

So now that I have taught you not to be afraid of Madam How, we will go and watch her at her work; and if we do not understand anything we see, we will ask her questions. She will always show us one of her lesson books if we give her time. And if we have to wait some time for her answer, you need not fear catching cold, though it is November; for she keeps her lesson books scattered about in strange places, and we may have to walk up and down that hill more than once before we can make out how she makes the glen.

Well—how was the glen made? You shall guess it if you like, and I will guess too. You think, perhaps, that an earthquake opened it?

My dear child, we must look before we guess. Then, after we have looked a little, and got some grounds for guessing, then we may guess. And you have no ground for supposing there ever was an earthquake here strong enough to open that glen. There may have been one: but we must guess from what we do know, and not from what we do not.

THE WATERFALL

Guess again. Perhaps it was there always, from the beginning of the world? My dear child, you have no proof of that either. Everything round you is changing in shape daily and hourly, as you will find out the longer you live; and therefore it is most reasonable to suppose that this glen has changed its shape, as everything else on earth has done. Besides, I told you not that Madam How had made the glen, but that she was making it, and as yet has only half finished. That is my first guess; and my next guess is that water is making the glen—water, and nothing else.

You open your young eyes. And I do not blame you. I looked at this very glen for fifteen years before I made that guess; and I have looked at it some ten years since, to make sure that my guess held good. For man after all is very blind, my dear boy, and very stupid, and cannot see what lies under his own feet all day long; and if Lady Why, and He whom Lady Why obeys, were not very patient and gentle with mankind, they would have perished off the face of the earth long ago, simply from their own stupidity. I, at least, was very stupid in this case, for I had my head full of earthquakes, and convulsions of nature, and all sorts of prodigies which never happened to this glen; and so, while I was trying to find what was not there, I of course found nothing. But when I put them all out of my head, and began to look for what was there, I found it at once; and lo and behold! I had seen it a thousand times before, and yet never learnt anything from it, like a stupid man as I was; though what I learnt you may learn as easily as I did.

And what did I find?

The pond at the bottom of the glen.

You know that pond, of course? You don't need to go there? Very well. Then if you do, do not you know also that the pond is always filling up with sand and mud; and that though we clean it out every three or four years, it always fills again? Now where does that sand and mud come from?

Down that stream, of course, which runs out of this bog. You see it coming down every time there is a flood, and the stream fouls.

Very well. Then, said Madam How to me, as soon as I recollected that, "Don't you see, you stupid man, that the stream has made the glen, and the earth which runs down the stream was all once part of the hill on which you stand." I confess I was very much ashamed of myself when she said that. For that is the history of the whole mystery. Madam How is digging away with her soft spade, water. She has a harder spade, or rather plough, the strongest and most terrible of all ploughs; but that, I am glad to say, she has laid by in England here.

Water? But water is too simple a thing to have dug out all this great glen.

My dear child, the most wonderful part of Madam How's work is, that she does such great things and so many different things, with one and the same tool, which looks to you so simple, though it really is not so. Water, for instance, is not a simple thing, but most complicated; and we might spend hours in talking about water, without having come to the end of its wonders. Still Madam How is a great economist, and never wastes her materials. She is like the sailor who boasted (only she never boasts) that, if he had but a long life and a strong knife, he would build St. Paul's Cathedral before he was done. And Madam How has a very long life, and plenty of time; and one of the strongest of all her tools is water. Now if you will stoop down and look into the heather, I will show you how she is digging out the glen with this very mist which is hanging about our feet. At least, so I guess.

For see how the mist clings to the points of the heather leaves, and makes drops. If the hot sun came out the drops would dry, and they would vanish into the air in light warm steam. But now that it is dark and cold they drip, or run down the heather-stems, to the ground. And whither do they go then? Whither will the water go,—hundreds of gallons of it perhaps,—which has dripped and run through the heather in this single day? It will sink into the ground, you know. And then what will become of it? Madam How will use it as an underground spade, just as she uses the rain (at least, when it rains too hard, and therefore the rain runs off the moor instead of sinking into it) as a spade above ground.

THE RAVINE

Now come to the edge of the glen, and I will show you the mist that fell yesterday, perhaps, coming out of the ground again, and hard at work.

You know of what an odd, and indeed of what a pretty form all these glens are. How the flat moor ends suddenly in a steep rounded bank, almost like the crest of a wave—ready like a wave-crest to fall over, and as you know, falling over sometimes, bit by bit, where the soil is bare.

Oh, yes; you are very fond of those banks. It is "awfully jolly," as you say, scrambling up and down them, in the deep heath and fern; besides, there are plenty of rabbit-holes there, because they are all sand; while there are no rabbit-holes on the flat above, because it is all gravel.

Yes; you know all about it: but you know, too, that you must not go too far down these banks, much less roll down them, because there is almost certain to be a bog at the bottom, lying upon a gentle slope; and there you get wet through.

All round these hills, from here to Aldershot in one direction, and from here to Windsor in another, you see the same shaped glens; the wave-crest along their top, and at the foot of the crest a line of springs which run out over the slopes, or well up through them in deep sand-galls, as you call them—shaking quagmires which are sometimes deep enough to swallow up a horse, and which you love to dance upon in summer time. Now the water of all these springs is nothing but the rain, and mist, and dew, which has sunk down first through the peaty soil, and then through the gravel and sand, and there has stopped. And why? Because under the gravel (about which I will tell you a strange story one day) and under the sand, which is what the geologists call the Upper Bagshot sand, there is an entirely different set of beds, which geologists call the Bracklesham beds, from a place near the New Forest; and in those beds there is a vein of clay, and through that clay the water cannot get, as you have seen yourself when we dug it out in the field below to puddle the pond-head; and very good fun you thought it, and a very pretty mess you made of yourself. Well: because the water cannot get though this clay, and must go somewhere, it runs out continually along the top of the clay, and as it runs undermines the bank, and brings down sand and gravel continually for the next shower to wash into the stream below.

Now think for one moment how wonderful it is that the shape of these glens, of which you are so fond, was settled by the particular order in which Madam How laid down the gravel and sand and mud at the bottom of the sea, ages and ages ago. This is what I told you, that the least thing that Madam How does to-day may take effect hundreds and thousands of years hence.

But I must tell you I think there was a time when this glen was of a very different shape from what it is now; and I dare say, according to

your notions, of a much prettier shape. It was once just like one of
those Chines which we used to see at Bournemouth. You recollect
them? How there was a narrow gap in the cliff of striped sands and
gravels; and out of the mouth of that gap, only a few feet across, there
poured down a great slope of mud and sand the shape of half a bun,
some wet and some dry, up which we used to scramble and get into the
Chine, and call the Chine what it was in the truest sense, Fairyland.
You recollect how it was all eaten out into mountain ranges, pinnacles,
steep cliffs of white, and yellow, and pink, standing up against the clear
blue sky; till we agreed that, putting aside the difference of size, they
were as beautiful and grand as any Alps we had ever seen in pictures.
And how we saw (for there could be no mistake about it there) that the
Chine was being hollowed out by the springs which broke out high up
the cliff, and by the rain which wore the sand into furrowed pinnacles
and peaks. You recollect the beautiful place, and how, when we looked
back down it we saw between the miniature mountain walls the bright
blue sea, and heard it murmur on the sands outside. So I verily believe
we might have done, if we had stood somewhere at the bottom of this
glen thousands of years ago. We should have seen the sea in front of us;
or rather, an arm of the sea; for Finchampstead ridges opposite, instead
of being covered with farms, and woodlands, and purple heath above,
would have been steep cliffs of sand and clay, just like those you see at
Bournemouth now; and—what would have spoilt somewhat the beauty
of the sight—along the shores there would have floated, at least in
winter, great blocks and floes of ice, such as you might have seen in the
tideway at King's Lynn the winter before last, growling and crashing,
grubbing and ploughing the sand, and the gravel, and the mud, and
sweeping them away into seas towards the North, which are now all
fruitful land. That may seem to you like a dream: yet it is true; and
some day, when we have another talk with Madam How, I will show
even a child like you that it was true.

But what could change a beautiful Chine like that at Bournemouth
into a wide sloping glen like this of Bracknell's Bottom, with a wood
like Coombs', many acres large, in the middle of it? Well now, think. It
is a capital plan for finding out Madam How's secrets, to see what she
might do in one place, and explain by it what she has done in another.
Suppose now, Madam How had orders to lift up the whole coast of
Bournemouth only twenty or even ten feet higher out of the sea than it
is now. She could do that easily enough, for she has been doing so on
the coast of South America for ages; she has been doing so this very
summer in what hasty people would call a hasty, and violent, and
ruthless way; though I shall not say so, for I believe that Lady Why
knows best. She is doing so now steadily on the west coast of Norway,
which is rising quietly—all that vast range of mountain wall and iron-
bound cliff—at the rate of some four feet in a hundred years, without

making the least noise or confusion, or even causing an extra ripple on the sea; so light and gentle, when she will, can Madam How's strong finger be.

Now, if the mouth of that Chine at Bournemouth was lifted twenty feet out of the sea, one thing would happen,—that the high tide would not come up any longer, and wash away the cake of dirt at the entrance, as we saw it do so often. But if the mud stopped there, the mud behind it would come down more slowly, and lodge inside more and more, till the Chine was half filled-up, and only the upper part of the cliffs continue to be eaten away, above the level where the springs ran out. So gradually the Chine, instead of being deep and narrow, would become broad and shallow; and instead of hollowing itself rapidly after every shower of rain, as you saw the Chine at Bournemouth doing, would hollow itself out slowly, as this glen is doing now. And one thing more would happen,—when the sea ceased to gnaw at the foot of the cliffs outside, and to carry away every stone and grain of sand which fell from them, the cliffs would very soon cease to be cliffs; the rain and the frost would still crumble them down, but the dirt that fell would lie at their feet, and gradually make a slope of dry land, far out where the shallow sea had been; and their tops, instead of being steep as now, would become smooth and rounded; and so at last, instead of two sharp walls of cliff at the Chine's mouth, you might have—just what you have here at the mouth of this glen,—our Mount and the Warren Hill,—long slopes with sheets of drifted gravel and sand at their feet, stretching down into what was once an icy sea, and is now the Vale of Blackwater. And this I really believe Madam How has done simply by lifting Hartford Bridge Flat a few more feet out of the sea, and leaving the rest to her trusty tool, the water in the sky.

That is my guess: and I think it is a good guess, because I have asked Madam How a hundred different questions about it in the last ten years, and she always answered them in the same way, saying, "Water, water, you stupid man." But I do not want you merely to depend on what I say. If you want to understand Madam How, you must ask her questions yourself, and make up your mind yourself like a man, instead of taking things at hearsay or second-hand, like the vulgar. Mind, by "the vulgar" I do not mean poor people: I mean ignorant and uneducated people, who do not use their brains rightly, though they may be fine ladies, kings, or popes. The Bible says, "Prove all things: hold fast that which is good." So do you prove my guess, and if it proves good, hold it fast.

And how can I do that?

First, by direct experiment, as it is called. In plain English—go home and make a little Hartford Bridge Flat in the stable-yard; and then ask Mrs. How if she will not make a glen in it like this glen here. We will go home and try that. We will make a great flat cake of clay, and put upon it a cap of sand; and then we will rain upon it out of a watering-pot; and see if Mrs. How does not begin soon to make a glen in the side of the heap, just like those on Hartford Bridge Flat. I believe she will; and certainly, if she does, it will be a fresh proof that my guess is right. And then we will see whether water will not make glens of a different shape than these, if it run over soils of a different kind. We will make a Hartford Bridge Flat turned upside down—a cake of sand with a cap of clay on the top; and we will rain on that out of our watering-pot, and see what sort of glens we make then. I can guess what they will be like, because I have seen them—steep overhanging cliffs, with very narrow gullies down them: but you shall try for yourself, and make up your mind whether you think me right or wrong. Meanwhile, remember that those gullies too will have been made by water.

And there is another way of "verifying my theory," as it is called; in plain English, seeing if my guess holds good; that is, to look at other valleys—not merely the valleys round here, but valleys in clay, in chalk, in limestone, in the hard slate rock such as you saw in Devonshire—and see whether my guess does not hold good about them too; whether all of them, deep or shallow, broad or narrow, rock or earth, may not have been all hollowed out by running water. I am sure if you would do this you would find something to amuse you, and something to instruct you, whenever you wish. I know that I do. To me the longest railroad journey, instead of being stupid, is like continually turning over the leaves of a wonderful book, or looking at wonderful pictures of old worlds which were made and unmade thousands of years ago. For I keep looking, not only at the railway cuttings, where the bones of the old worlds are laid bare, but at the surface of the ground; at the plains and downs, banks and knolls, hills and mountains; and continually asking Mrs. How what gave them each its shape: and I will soon teach you to do the same. When you do, I tell you fairly her answer will be in almost every case, "Running water." Either water running when soft, as it usually is; or water running when it is hard—in plain words, moving ice.

About that moving ice, which is Mrs. How's stronger spade, I will tell you some other time; and show you, too, the marks of it in every gravel pit about here. But now, I see, you want to ask a question; and what is it?

Do I mean to say that water has made great valleys, such as you have seen paintings and photographs of,—valleys thousands of feet deep, among mountains thousands of feet high?

Yes, I do. But, as I said before, I do not like you to take my word upon trust. When you are older you shall go to the mountains, and you shall judge for yourself. Still, I must say that I never saw a valley, however deep, or a cliff, however high, which had not been scooped out by water; and that even the mountain-tops which stand up miles aloft in jagged peaks and pinnacles against the sky were cut out at first, and are being cut and sharpened still, by little else save water, soft and hard; that is, by rain, frost, and ice.

THE MATTERHORN

Water, and nothing else, has sawn out such a chasm as that through which the ships run up to Bristol, between Leigh Wood and St. Vincent's Rocks. Water, and nothing else, has shaped those peaks of the Matterhorn, or the Weisshorn, or the Pic du Midi of the Pyrenees, of which you have seen sketches and photographs. Just so water might saw out Hartford Bridge Flat, if it had time enough, into a labyrinth of valleys, and hills, and peaks standing alone; as it has done already by Ambarrow, and Edgbarrow, and the Folly Hill on the other side of the vale.

I see you are astonished at the notion that water can make Alps. But it was just because I knew you would be astonished at Madam How's doing so great a thing with so simple a tool, that I began by showing you how she was doing the same thing in a small way here upon these flats. For the safest way to learn Madam How's methods is to watch her at work in little corners at commonplace business, which will not astonish or frighten us, nor put huge hasty guesses and dreams into our heads. Sir Isaac Newton, some will tell you, found out the great law of gravitation, which holds true of all the suns and stars in heaven, by watching an apple fall: and even if he did not find it out so, he found it out, we know, by careful thinking over the plain and commonplace fact, that things have weight. So do you be humble and patient, and watch Madam How at work on little things. For that is the way to see her at work upon all space and time.

What? you have a question more to ask?

Oh! I talked about Madam How lifting up Hartford Bridge Flat. How could she do that? My dear child, that is a long story, and I must tell it you some other time. Meanwhile, did you ever see the lid of a kettle rise up and shake when the water inside boiled? Of course; and of course, too, remember that Madam How must have done it. Then think over between this and our next talk, what that can possibly have to do with her lifting up Hartford Bridge Flat. But you have been longing, perhaps, all this time to hear more about Lady Why, and why she set Madam How to make Bracknell's Bottom.

My dear child, the only answer I dare give to that is: Whatever other purposes she may have made it for, she made it at least for this— that you and I should come to it this day, and look at, and talk over it, and become thereby wiser and more earnest, and we will hope more humble and better people. Whatever else Lady Why may wish or not wish, this she wishes always, to make all men wise and all men good. For what is written of her whom, as in a parable, I have called Lady Why?

"The Lord possessed me in the beginning of His way, before His works of old.

"I was set up from everlasting, from the beginning, or ever the earth was.

"When there were no depths, I was brought forth; when there were no fountains abounding with water.

"Before the mountains were settled, before the hills was I brought forth:

"While as yet He had not made the earth, nor the fields, nor the highest part of the dust of the world.

"When He prepared the heavens, I was there: when He set a compass upon the face of the depth:

"When He established the clouds above: when He strengthened the fountains of the deep:

"When He gave to the sea His decree, that the waters should not pass His commandment: when He appointed the foundations of the earth:

"Then I was by Him, as one brought up with Him: and I was daily His delight, rejoicing always before Him:

"Rejoicing in the habitable part of His earth; and my delights were with the sons of men.

"Now therefore hearken unto me, O ye children: for blessed are they that keep my ways."

That we can say, for it has been said for us already. But beyond that we can say, and need say, very little. We were not there, as we read in the Book of Job, when God laid the foundations of the earth. "We see," says St. Paul, "as in a glass darkly, and only know in part." "For who," he asks again, "has known the mind of the Lord, or who hath been His counsellor? . . . For of Him, and through Him, and to Him, are all things: to whom be glory for ever and ever. Amen." Therefore we must not rashly say, this or that is Why a thing has happened; nor invent what are called "final causes," which are not Lady Why herself, but only our little notions of what Lady Why has done, or rather what we should have done if we had been in her place. It is not, indeed, by thinking that we shall find out anything about Lady Why. She speaks not to our eyes or to our brains, like Madam How, but to that inner part of us which we call our hearts and spirits, and which will endure when eyes and brain are turned again to dust. If your heart be pure and sober, gentle and truthful, then Lady Why speaks to you without words, and tells you things which Madam How and all her pupils, the men of science, can never tell. When you lie, it may be, on a painful sick-bed, but with your mother's hand in yours; when you sit by her, looking up into her loving eyes; when you gaze out towards the setting sun, and fancy golden capes and islands in the clouds, and seas and lakes in the blue sky, and the infinite rest and peace of the far west sends rest and peace into your young heart, till you sit silent and happy, you know not why; when sweet music fills your heart with noble and tender instincts which need no thoughts or words; ay, even when you watch the raging thunderstorm, and feel it to be, in spite of its great awfulness, so beautiful that you cannot turn your eyes away: at such times as these Lady Why is speaking to your soul of souls, and saying, "My child, this world is a new place, and strange, and often terrible: but be not afraid. All will come right at last. Rest will conquer Restlessness; Faith will conquer Fear; Order will conquer Disorder; Health will conquer Sickness; Joy will conquer Sorrow; Pleasure will conquer Pain; Life will conquer Death; Right will conquer Wrong. All will be well at last. Keep your soul and body pure, humble, busy, pious—in one word, be good: and ere you die, or after you die, you may have some glimpse of Me, the Everlasting Why: and hear with the ears, not of your body but of your spirit, men and all rational beings, plants and animals, ay, the very stones beneath your feet, the clouds above your head, the planets and the suns away in farthest space, singing eternally,

"'Thou art worthy, O Lord, to receive glory and honour and power, for Thou hast created all things, and for Thy pleasure they are and were created.'"

Chapter II. Earthquakes

So? You have been looking at that beautiful drawing of the ruin of Arica in the *Illustrated London News*: and it has puzzled you and made you sad. You want to know why God killed all those people—mothers among them, too, and little children?

Alas, my dear child! who am I that I should answer you that?

Have you done wrong in asking me? No, my dear child; no. You have asked me because you are a human being and a child of God, and not merely a cleverer sort of animal, an ape who can read and write and cast accounts. Therefore it is that you cannot be content, and ought not to be content, with asking how things happen, but must go on to ask why. You cannot be content with knowing the causes of things; and if you knew all the natural science that ever was or ever will be known to men, that would not satisfy you; for it would only tell you the *causes* of things, while your souls want to know the *reasons* of things besides; and though I may not be able to tell you the reasons of things, or show you aught but a tiny glimpse here and there of that which I called the other day the glory of Lady Why, yet I believe that somehow, somewhen, somewhere, you will learn something of the reason of things. For that thirst to know *why* was put into the hearts of little children by God Himself; and I believe that God would never have given them that thirst if He had not meant to satisfy it.

ARICA AFTER THE EARTHQUAKE

There—you do not understand me. I trust that you will understand me some day. Meanwhile, I think—I only say I *think*—you know I told you how humble we must be whenever we speak of Lady Why—that we may guess at something like a good reason for the terrible earthquakes in South America. I do not wish to be hard upon poor people in great affliction: but I cannot help thinking that they have been doing for hundreds of years past something very like what the Bible calls "tempting God"—staking their property and their lives upon the chances of no earthquakes coming, while they ought to have known that an earthquake might come any day. They have fulfilled (and little thought I that it would be fulfilled so soon) the parable that I told you once, of the nation of the Do-as-you-likes, who lived careless and happy at the foot of the burning mountain, and would not be warned by the smoke that came out of the top, or by the slag and cinders which lay all about them; till the mountain blew up, and destroyed them miserably.

Then I think that they ought to have expected an earthquake.

Well—it is not for us to judge any one, especially if they live in a part of the world in which we have not been ourselves. But I think that we know, and that they ought to have known, enough about earthquakes to have been more prudent than they have been for many a year. At least we will hope that, though they would not learn their lesson till this year, they will learn it now, and will listen to the message which I think Madam How has brought them, spoken in a voice of thunder, and written in letters of flame.

And what is that?

My dear child, if the landlord of our house was in the habit of pulling the roof down upon our heads, and putting gunpowder under the foundations to blow us up, do you not think we should know what he meant, even though he never spoke a word? He would be very wrong in behaving so, of course: but one thing would be certain,—that he did not intend us to live in his house any longer if he could help it; and was giving us, in a very rough fashion, notice to quit. And so it seems to me that these poor Spanish Americans have received from the Landlord of all landlords, who can do no wrong, such a notice to quit as perhaps no people ever had before; which says to them in unmistakable words, "You must leave this country: or perish." And I believe that that message, like all Lady Why's messages, is at heart a merciful and loving one; that if these Spaniards would leave the western coast of Peru, and cross the Andes into the green forests of the eastern side of their own land, they might not only live free from earthquakes, but (if they would only be good and industrious) become a great, rich, and happy nation, instead of the idle, and useless, and I am afraid not over good, people which they have been. For in that eastern part of their own land God's gifts are waiting for them, in a paradise such as I can neither describe nor you conceive;—precious woods, fruits, drugs, and what not—boundless wealth, in one word—waiting for them to send it all down the waters of the mighty river Amazon, enriching us here in the Old World, and enriching themselves there in the New. If they would only go and use these gifts of God, instead of neglecting them as they have been doing for now three hundred years, they would be a blessing to the earth, instead of being—that which they have been.

God grant, my dear child, that these poor people may take the warning that has been sent to them; "The voice of God revealed in facts," as the great Lord Bacon would have called it, and see not only that God has bidden them leave the place where they are now, but has prepared for them, in their own land, a home a thousand times better than that in which they now live.

But you ask, How ought they to have known that an earthquake would come?

Well, to make you understand that, we must talk a little about earthquakes, and what makes them; and in order to find out that, let us try the very simplest cause of which we can think. That is the wise and scientific plan.

Now, whatever makes these earthquakes must be enormously strong; that is certain. And what is the strongest thing you know of in the world? Think . . .

Gunpowder?

Well, gunpowder is strong sometimes: but not always. You may carry it in a flask, or in your hand, and then it is weak enough. It only becomes strong by being turned into gas and steam. But steam is always strong. And if you look at a railway engine, still more if you had ever seen—which God forbid you should—a boiler explosion, you would agree with me, that the strongest thing we know of in the world is steam.

Now I think that we can explain almost, if not quite, all that we know about earthquakes, if we believe that on the whole they are caused by steam and other gases expanding, that is, spreading out, with wonderful quickness and strength. Of course there must be something to make them expand, and that is *heat*. But we will not talk of that yet.

Now do you remember that riddle which I put to you the other day?—"What had the rattling of the lid of the kettle to do with Hartford Bridge Flat being lifted out of the ancient sea?"

The answer to the riddle, I believe, is—Steam has done both. The lid of the kettle rattles, because the expanding steam escapes in little jets, and so causes a *lid-quake*. Now suppose that there was steam under the earth trying to escape, and the earth in one place was loose and yet hard, as the lid of the kettle is loose and yet hard, with cracks in it, it may be, like the crack between the edge of the lid and the edge of the kettle itself: might not the steam try to escape through the cracks, and rattle the surface of the earth, and so cause an *earthquake*?

So the steam would escape generally easily, and would only make a passing rattle, like the earthquake of which the famous jester Charles Selwyn said that it was quite a young one, so tame that you might have stroked it; like that which I myself once felt in the Pyrenees, which gave me very solemn thoughts after a while, though at first I did nothing but laugh at it; and I will tell you why.

I was travelling in the Pyrenees; and I came one evening to the loveliest spot—a glen, or rather a vast crack in the mountains, so narrow that there was no room for anything at the bottom of it, save a torrent roaring between walls of polished rock. High above the torrent the road was cut out among the cliffs, and above the road rose more cliffs, with great black cavern mouths, hundreds of feet above our heads, out of each of which poured in foaming waterfalls streams large enough to turn a mill, and above them mountains piled on mountains, all covered with woods of box, which smelt rich and hot and musky in the warm spring air. Among the box-trees and fallen boulders grew hepaticas, blue and white and red, such as you see in the garden; and little stars of gentian, more azure than the azure sky. But out of the box-woods above rose giant silver firs, clothing the cliffs and glens with tall black spires, till they stood out at last in a jagged saw-edge against the purple evening sky, along the mountain ranges, thousands of feet aloft; and beyond them again, at the head of the valley, rose vast cones of virgin snow, miles away in reality, but looking so brilliant and so near that one fancied at the first moment that one could have touched them with one's hand. Snow-white they stood, the glorious things, seven thousand feet into the air; and I watched their beautiful white sides turn rose-colour in the evening sun, and when he set, fade into dull cold gray, till the bright moon came out to light them up once more. When I was tired of wondering and admiring, I went into bed; and there I had a dream—such a dream as Alice had when she went into Wonderland—such a dream as I dare say you may have had ere now. Some noise or stir puts into your fancy as you sleep a whole long dream to account for it; and yet that dream, which seems to you to be hours long, has not taken up a second of time; for the very same noise which begins the dream, wakes you at the end of it: and so it was with me. I dreamed that some English people had come into the hotel where I was, and were sleeping in the room underneath me; and that they had quarrelled and fought, and broke their bed down with a tremendous crash, and that I must get up, and stop the fight; and at that moment I woke and heard coming up the valley from the north such a roar as I never heard before or since; as if a hundred railway trains were rolling underground; and just as it passed under my bed there was a tremendous thump, and I jumped out of bed quicker than I ever did in my life, and heard the roaring sound die away as it rolled up the valley towards the peaks of snow. Still I had in my head this notion of the Englishmen fighting in the room below. But then I recollected that no Englishmen had come in the night before, and that I had been in the room below, and that there was no bed in it. Then I opened my window—a woman screamed, a dog barked, some cocks and hens cackled in a very disturbed humour, and then I could hear nothing but the roaring of the torrent a hundred feet below. And then it flashed across me what all the noise was about; and I burst out laughing and said "It is only an earthquake," and went to bed again.

Next morning I inquired whether any one had heard a noise. No, nobody had heard anything. And the driver who had brought me up the valley only winked, but did not choose to speak. At last at breakfast I asked the pretty little maid who waited what was the meaning of the noise I heard in the night, and she answered, to my intense amusement, "Ah! bah! ce n'etait qu'un tremblement de terre; il y en a ici toutes les six semaines." Now the secret was out. The little maid, I found, came from the lowland far away, and did not mind telling the truth: but the good people of the place were afraid to let out that they had earthquakes every six weeks, for fear of frightening visitors away: and because they were really very good people, and very kind to me, I shall not tell you what the name of the place is.

Of course after that I could do no less than ask Madam How, very civilly, how she made earthquakes in that particular place, hundreds of miles away from any burning mountain? And this was the answer I *thought* she gave, though I am not so conceited as to say I am sure.

As I had come up the valley I had seen that the cliffs were all beautiful gray limestone marble; but just at this place they were replaced by granite, such as you may see in London Bridge or at Aberdeen. I do not mean that the limestone changed to granite, but that the granite had risen up out of the bottom of the valley, and had carried the limestone (I suppose) up on its back hundreds of feet into the air. Those caves with the waterfalls pouring from their mouths were all on one level, at the top of the granite, and the bottom of the limestone. That was to be expected; for, as I will explain to you some day, water can make caves easily in limestone: but never, I think, in granite. But I knew that besides these cold springs which came out of the caves, there were hot springs also, full of curious chemical salts, just below the very house where I was in. And when I went to look at them, I found that they came out of the rock just where the limestone and the granite joined. "Ah," I said, "now I think I have Madam How's answer. The lid of one of her great steam boilers is rather shaky and cracked just here, because the granite has broken and torn the limestone as it lifted it up; and here is the hot water out of the boiler actually oozing out of the crack; and the earthquake I heard last night was simply the steam rumbling and thumping inside, and trying to get out."

And then, my dear child, I fell into a more serious mood. I said to myself, "If that stream had been a little, only a little stronger, or if the rock above it had been only a little weaker, it would have been no laughing matter then; the village might have been shaken to the ground; the rocks hurled into the torrent; jets of steam and of hot water, mixed, it may be, with deadly gases, have roared out of the riven ground; that might have happened here, in short, which has happened and happens still in a hundred places in the world, whenever the rocks are too weak to stand the pressure of the steam below, and the solid earth bursts as an engine boiler bursts when the steam within it is too strong." And when those thoughts came into my mind, I was in no humour to jest any more about "young earthquakes," or "Madam How's boilers;" but rather to say with the wise man of old, "It is of the Lord's mercies that we are not consumed."

Most strange, most terrible also, are the tricks which this underground steam plays. It will make the ground, which seems to us so hard and firm, roll and rock in waves, till people are sea-sick, as on board a ship; and that rocking motion (which is the most common) will often, when it is but slight, set the bells ringing in the steeples, or make the furniture, and things on shelves, jump about quaintly enough. It will make trees bend to and fro, as if a wind was blowing through them; open doors suddenly, and shut them again with a slam; make the timbers of the floors and roofs creak, as they do in a ship at sea; or give men such frights as one of the dock-keepers at Liverpool got in the earthquake in 1863, when his watchbox rocked so, that he thought some one was going to pitch him over into the dock. But these are only little hints and warnings of what it can do. When it is strong enough, it will rock down houses and churches into heaps of ruins, or, if it leaves them standing, crack them from top to bottom, so that they must be pulled down and rebuilt.

You saw those pictures of the ruins of Arica, about which our talk began; and from them you can guess well enough for yourself what a town looks like which has been ruined by an earthquake. Of the misery and the horror which follow such a ruin I will not talk to you, nor darken your young spirit with sad thoughts which grown people must face, and ought to face. But the strangeness of some of the tricks which the earthquake shocks play is hardly to be explained, even by scientific men. Sometimes, it would seem, the force runs round, making the solid ground eddy, as water eddies in a brook. For it will make straight rows of trees crooked; it will twist whole walls round—or rather the ground on which the walls stand—without throwing them down; it will shift the stones of a pillar one on the other sideways, as if a giant had been trying to spin it like a teetotum, and so screwed it half in pieces. There is a story told by a wise man, who saw the place himself, of the whole furniture of one house being hurled away by an earthquake, and buried under the ruins of another house; and of things carried hundreds of yards off, so that the neighbours went to law to settle who was the true owner of them. Sometimes, again, the shock seems to come neither horizontally in waves, nor circularly in eddies, but vertically, that is, straight up from below; and then things—and people, alas! sometimes—are thrown up off the earth high into the air, just as things spring up off the table if you strike it smartly enough underneath. By that same law (for there is a law for every sort of motion) it is that the earthquake shock sometimes hurls great rocks off a cliff into the valley below. The shock runs through the mountain till it comes to the cliff at the end of it; and then the face of the cliff, if it be at all loose, flies off into the air. You may see the very same thing happen, if you will put marbles or billiard-balls in a row touching each other, and strike the one nearest you smartly in the line of the row. All the balls stand still, except the last one, and that flies off. The shock, like the earthquake shock, has run through them all; but only the end one, which had nothing beyond it but soft air, has been moved; and when you grow old, and learn mathematics, you will know the law of motion according to which that happens, and learn to apply what the billiard-balls have taught you, to explain the wonders of an earthquake. For in this case, as in so many more, you must watch Madam How at work on little and common things, to find out how she works in great and rare ones. That is why Solomon says that "a fool's eyes are in the ends of the earth," because he is always looking out for strange things which he has not seen, and which he could not understand if he saw; instead of looking at the petty commonplace matters which are about his feet all day long, and getting from them sound knowledge, and the art of getting more sound knowledge still.

Another terrible destruction which the earthquake brings, when it is close to the seaside, is the wash of a great sea wave, such as swept in last year upon the island of St. Thomas, in the West Indies; such as swept in upon the coast of Peru this year. The sea moans, and sinks back, leaving the shore dry; and then comes in from the offing a mighty wall of water, as high as, or higher than, many a tall house; sweeps far inland, washing away quays and houses, and carrying great ships in with it; and then sweeps back again, leaving the ships high and dry, as ships were left in Peru this year.

Now, how is that wave made? Let us think. Perhaps in many ways. But two of them I will tell you as simply as I can, because they seem the most likely, and probably the most common.

Suppose, as the earthquake shock ran on, making the earth under the sea heave and fall in long earth-waves, the sea-bottom sank down. Then the water on it would sink down too, and leave the shore dry; till the sea-bottom rose again, and hurled the water up again against the land. This is one way of explaining it, and it may be true. For certain it is, that earthquakes do move the bottom of the sea; and certain, too, that they move the water of the sea also, and with tremendous force. For ships at sea during an earthquake feel such a blow from it (though it does them no harm) that the sailors often rush upon deck fancying that they have struck upon a rock; and the force which could give a ship, floating in water, such a blow as that, would be strong enough to hurl thousands of tons of water up the beach, and on to the land.

But there is another way of accounting for this great sea wave, which I fancy comes true sometimes.

Suppose you put an empty india-rubber ball into water, and then blow into it through a pipe. Of course, you know, as the ball filled, the upper side of it would rise out of the water. Now, suppose there were a party of little ants moving about upon that ball, and fancying it a great island, or perhaps the whole world—what would they think of the ball's filling and growing bigger?

If they could see the sides of the basin or tub in which the ball was, and were sure that they did not move, then they would soon judge by them that they themselves were moving, and that the ball was rising out of the water. But if the ants were so short-sighted that they could not see the sides of the basin, they would be apt to make a mistake, because they would then be like men on an island out of sight of any other land. Then it would be impossible further to tell whether they were moving up, or whether the water was moving down; whether their ball was rising out of the water, or the water was sinking away from the ball. They would probably say, "The water is sinking and leaving the ball dry."

Do you understand that? Then think what would happen if you pricked a hole in the ball. The air inside would come hissing out, and the ball would sink again into the water. But the ants would probably fancy the very opposite. Their little heads would be full of the notion that the ball was solid and could not move, just as our heads are full of the notion that the earth is solid and cannot move; and they would say, "Ah! here is the water rising again." Just so, I believe, when the sea seems to ebb away during the earthquake, the land is really being raised out of the sea, hundreds of miles of coast, perhaps, or a whole island, at once, by the force of the steam and gas imprisoned under the ground. That steam stretches and strains the solid rocks below, till they can bear no more, and snap, and crack, with frightful roar and clang; then out of holes and chasms in the ground rush steam, gases—often foul and poisonous ones—hot water, mud, flame, strange stones—all signs that the great boiler down below has burst at last.

Then the strain is eased. The earth sinks together again, as the ball did when it was pricked; and sinks lower, perhaps, than it was before: and back rushes the sea, which the earth had thrust away while it rose, and sweeps in, destroying all before it.

Of course, there is a great deal more to be said about all this: but I have no time to tell you now. You will read it, I hope, for yourselves when you grow up, in the writings of far wiser men than I. Or perhaps you may feel for yourselves in foreign lands the actual shock of a great earthquake, or see its work fresh done around you. And if ever that happens, and you be preserved during the danger, you will learn for yourself, I trust, more about earthquakes than I can teach you, if you will only bear in mind the simple general rules for understanding the "how" of them which I have given you here.

But you do not seem satisfied yet? What is it that you want to know?

Oh! There was an earthquake here in England the other night, while you were asleep; and that seems to you too near to be pleasant. Will there ever be earthquakes in England which will throw houses down, and bury people in the ruins?

My dear child, I think you may set your heart at rest upon that point. As far as the history of England goes back, and that is more than a thousand years, there is no account of any earthquake which has done any serious damage, or killed, I believe, a single human being. The little earthquakes which are sometimes felt in England run generally up one line of country, from Devonshire through Wales, and up the Severn valley into Cheshire and Lancashire, and the south-west of Scotland; and they are felt more smartly there, I believe, because the rocks are harder there than here, and more tossed about by earthquakes which happened ages and ages ago, long before man lived on the earth. I will show you the work of these earthquakes some day, in the tilting and twisting of the layers of rock, and in the cracks (*faults*, as they are called) which run through them in different directions. I showed you some once, if you recollect, in the chalk cliff at Ramsgate—two set of cracks, sloping opposite ways, which I told you were made by two separate sets of earthquakes, long, long ago, perhaps while the chalk was still at the bottom of a deep sea. But even in the rocky parts of England the earthquake-force seems to have all but died out. Perhaps the crust of the earth has become too thick and solid there to be much shaken by the gases and steam below. In this eastern part of England, meanwhile, there is but little chance that an earthquake will ever do much harm, because the ground here, for thousands of feet down, is not hard and rocky, but soft—sands, clays, chalk, and sands again; clays, soft limestones, and clays again—which all act as buffers to deaden the earthquake shocks, and deaden too the earthquake noise.

And how?

Put your ear to one end of a soft bolster, and let some one hit the other end. You will hear hardly any noise, and will not feel the blow at all. Put your ear to one end of a hard piece of wood, and let some one hit the other. You will hear a smart tap; and perhaps feel a smart tap, too. When you are older, and learn the laws of sound, and of motion among the particles of bodies, you will know why. Meanwhile you may comfort yourself with the thought that Madam How has (doubtless by command of Lady Why) prepared a safe soft bed for this good people of Britain—not that they may lie and sleep on it, but work and till, plant and build and manufacture, and thrive in peace and comfort, we will trust and pray, for many a hundred years to come. All that the steam inside the earth is likely to do to us, is to raise parts of this island (as Hartford Bridge Flats were raised, ages ago, out of the old icy sea) so slowly, probably, that no man can tell whether they are rising or not. Or again, the steam-power may be even now dying out under our island, and letting parts of it sink slowly into the sea, as some wise friends of mine think that the fens in Norfolk and Cambridgeshire are sinking now. I have shown you where that kind of work has gone on in Norfolk; how the brow of Sandringham Hill was once a sea-cliff, and Dersingham Bog at its foot a shallow sea; and therefore that the land has risen there. How, again, at Hunstanton Station there is a beach of sea-shells twenty feet above high-water mark, showing that the land has risen there likewise. And how, farther north again, at Brancaster, there are forests of oak, and fir, and alder, with their roots still in the soil, far below high-water mark, and only uncovered at low tide; which is a plain sign that there the land has sunk. You surely recollect the sunken forest at Brancaster, and the beautiful shells we picked up in its gullies, and the millions of live Pholases boring into the clay and peat which once was firm dry land, fed over by giant oxen, and giant stags likewise, and perhaps by the mammoth himself, the great woolly elephant whose teeth the fishermen dredge up in the sea outside? You recollect that? Then remember that as that Norfolk shore has changed, so slowly but surely is the whole world changing around us. Hartford Bridge Flat here, for instance, how has it changed! Ages ago it was the gravelly bottom of a sea. Then the steam-power underground raised it up slowly, through long ages, till it became dry land. And ages hence, perhaps, it will have become a sea-bottom once more. Washed slowly by the rain, or sunk by the dying out of the steam-power underground, it will go down again to the place from whence it came. Seas will roll where we stand now, and new lands will rise where seas now roll. For all things on this earth, from the tiniest flower to the tallest mountain, change and change all day long. Every atom of matter moves perpetually; and nothing "continues in one stay." The solid-seeming earth on which you stand is but a heaving bubble, bursting ever and anon in this place and in that. Only above all, and through all, and with all, is One who does not move nor change, but is the same yesterday, to-day, and for ever. And on Him, my child, and not on this bubble of an earth, do you and I, and all mankind, depend.

But I have not yet told you why the Peruvians ought to have

expected an earthquake. True. I will tell you another time.

Chapter III. Volcanoes

You want to know why the Spaniards in Peru and Ecuador should have expected an earthquake.

Because they had had so many already. The shaking of the ground in their country had gone on perpetually, till they had almost ceased to care about it, always hoping that no very heavy shock would come; and being, now and then, terribly mistaken.

For instance, in the province of Quito, in the year 1797, from thirty to forty thousand people were killed at once by an earthquake. One would have thought that warning enough: but the warning was not taken: and now, this very year, thousands more have been killed in the very same country, in the very same way.

They might have expected as much. For their towns are built, most of them, close to volcanoes—some of the highest and most terrible in the world. And wherever there are volcanoes there will be earthquakes. You may have earthquakes without volcanoes, now and then; but volcanoes without earthquakes, seldom or never.

How does that come to pass? Does a volcano make earthquakes? No; we may rather say that earthquakes are trying to make volcanoes. For volcanoes are the holes which the steam underground has burst open that it may escape into the air above. They are the chimneys of the great blast-furnaces underground, in which Madam How pounds and melts up the old rocks, to make them into new ones, and spread them out over the land above.

And are there many volcanoes in the world? You have heard of Vesuvius, of course, in Italy; and Etna, in Sicily; and Hecla, in Iceland. And you have heard, too, of Kilauea, in the Sandwich Islands, and of Pele's Hair—the yellow threads of lava, like fine spun glass, which are blown from off its pools of fire, and which the Sandwich Islanders believed to be the hair of a goddess who lived in the crater;—and you have read, too, I hope, in Miss Yonge's *Book of Golden Deeds*, the noble story of the Christian chieftainess who, in order to persuade her subjects to become Christians also, went down into the crater and defied the goddess of the volcano, and came back unhurt and triumphant.

But if you look at the map, you will see that there are many, many more. Get Keith Johnston's Physical Atlas from the schoolroom—of course it is there (for a schoolroom without a physical atlas is like a needle without an eye)—and look at the map which is called "Phenomena of Volcanic Action."

You will see in it many red dots, which mark the volcanoes which are still burning: and black dots, which mark those which have been burning at some time or other, not very long ago, scattered about the world. Sometimes they are single, like the red dot at Otaheite, or at Easter Island in the Pacific. Sometimes the are in groups, or clusters, like the cluster at the Sandwich Islands, or in the Friendly Islands, or in New Zealand. And if we look in the Atlantic, we shall see four clusters: one in poor half-destroyed Iceland, in the far north, one in the Azores, one in the Canaries, and one in the Cape de Verds. And there is one dot in those Canaries which we must not overlook, for it is no other than the famous Peak of Teneriffe, a volcano which is hardly burnt out yet, and may burn up again any day, standing up out of the sea more than 12,000 feet high still, and once it must have been double that height. Some think that it is perhaps the true Mount Atlas, which the old Greeks named when first they ventured out of the Straits of Gibraltar down the coast of Africa, and saw the great peak far to the westward, with the clouds cutting off its top; and said that it was a mighty giant, the brother of the Evening Star, who held up the sky upon his shoulders, in the midst of the Fortunate Islands, the gardens of the daughter of the Evening Star, full of strange golden fruits; and that Perseus had turned him into stone, when he passed him with the Gorgon's Head.

But you will see, too, that most of these red and black dots run in crooked lines; and that many of the clusters run in lines likewise.

Look at one line: by far the largest on the earth. You will learn a good deal of geography from it.

The red dots begin at a place called the Terribles, on the east side of the Bay of Bengal. They run on, here and there, along the islands of Sumatra and Java, and through the Spice Islands; and at New Guinea the line of red dots forks. One branch runs south-east, through islands whose names you never heard, to the Friendly Islands, and to New Zealand. The other runs north, through the Philippines, through Japan, through Kamschatka; and then there is a little break of sea, between Asia and America: but beyond it, the red dots begin again in the Aleutian Islands, and then turn down the whole west coast of America, down from Mount Elias (in what was, till lately, Russian America) towards British Columbia. Then, after a long gap, there are one or two in Lower California (and we must not forget the terrible earthquake which has just shaken San Francisco, between those two last places); and when we come down to Mexico we find the red dots again plentiful, and only too plentiful; for they mark the great volcanic line of Mexico, of which you will read, I hope, some day, in Humboldt's works. But the line does not stop there. After the little gap of the Isthmus of Panama, it begins again in Quito, the very country which has just been shaken, and in which stand the huge volcanoes Chimborazo, Pasto, Antisana, Cotopaxi, Pichincha, Tunguragua,— smooth cones from 15,000 to 20,000 feet high, shining white with snow, till the heat inside melts it off, and leaves the cinders of which the peaks are made all black and ugly among the clouds, ready to burst in smoke and fire. South of them again, there is a long gap, and then another line of red dots—Arequiba, Chipicani, Gualatieri, Atacama,—as high as, or higher than those in Quito; and this, remember, is the other country which has just been shaken. On the sea-shore below those volcanoes stood the hapless city of Arica, whose ruins we saw in the picture. Then comes another gap; and then a line of more volcanoes in Chili, at the foot of which happened that fearful earthquake of 1835 (besides many more) of which you will read some day in that noble book *The Voyage of the Beagle*; and so the line of dots runs down to the southernmost point of America.

What a line we have traced! Long enough to go round the world if it were straight. A line of holes out of which steam, and heat, and cinders, and melted stones are rushing up, perpetually, in one place and another. Now the holes in this line which are near each other have certainly something to do with each other. For instance, when the earth shook the other day round the volcanoes of Quito, it shook also round the volcanoes of Peru, though they were 600 miles away. And there are many stories of earthquakes being felt, or awful underground thunder heard, while volcanoes were breaking out hundreds of miles away. I will give you a very curious instance of that.

If you look at the West Indies on the map, you will see a line of red dots runs through the Windward Islands: there are two volcanoes in them, one in Guadaloupe, and one in St. Vincent (I will tell you a curious story, presently, about that last), and little volcanoes (if they have ever been real volcanoes at all), which now only send out mud, in Trinidad. There the red dots stop: but then begins along the north coast of South America a line of mountain country called Cumana, and Caraccas, which has often been horribly shaken by earthquakes. Now once, when the volcano in St. Vincent began to pour out a vast stream of melted lava, a noise like thunder was heard underground, over thousands of square miles beyond those mountains, in the plains of Calabozo, and on the banks of the Apure, more than 600 miles away from the volcano,—a plain sign that there was something underground which joined them together, perhaps a long crack in the earth. Look for yourselves at the places, and you will see that (as Humboldt says) it is as strange as if an eruption of Mount Vesuvius was heard in the north of France.

So it seems as if these lines of volcanoes stood along cracks in the rind of the earth, through which the melted stuff inside was for ever trying to force its way; and that, as the crack got stopped up in one place by the melted stuff cooling and hardening again into stone, it was burst in another place, and a fresh volcano made, or an old one re-opened.

Now we can understand why earthquakes should be most common round volcanoes; and we can understand, too, why they would be worst before a volcano breaks out, because then the steam is trying to escape; and we can understand, too, why people who live near volcanoes are glad to see them blazing and spouting, because then they have hope that the steam has found its way out, and will not make earthquakes any more for a while. But still that is merely foolish speculation on chance. Volcanoes can never be trusted. No one knows when one will break out, or what it will do; and those who live close to them—as the city of Naples is close to Mount Vesuvius—must not be astonished if they are blown up or swallowed up, as that great and beautiful city of Naples may be without a warning, any day.

For what happened to that same Mount Vesuvius nearly 1800 years ago, in the old Roman times? For ages and ages it had been lying quiet, like any other hill. Beautiful cities were built at its foot, filled with people who were as handsome, and as comfortable, and (I am afraid) as wicked, as people ever were on earth. Fair gardens, vineyards, olive-yards, covered the mountain slopes. It was held to be one of the Paradises of the world. As for the mountain's being a burning mountain, who ever thought of that? To be sure, on the top of it was a great round crater, or cup, a mile or more across, and a few hundred yards deep. But that was all overgrown with bushes and wild vines, full of boars and deer. What sign of fire was there in that? To be sure, also, there was an ugly place below by the sea-shore, called the Phlegræn fields, where smoke and brimstone came out of the ground, and a lake called Avernus over which poisonous gases hung, and which (old stories told) was one of the mouths of the Nether Pit. But what of that? It had never harmed any one, and how could it harm them?

So they all lived on, merrily and happily enough, till, in the year A.D. 79 (that was eight years, you know, after the Emperor Titus destroyed Jerusalem), there was stationed in the Bay of Naples a Roman admiral, called Pliny, who was also a very studious and learned man, and author of a famous old book on natural history. He was staying on shore with his sister; and as he sat in his study she called him out to see a strange cloud which had been hanging for some time over the top of Mount Vesuvius. It was in shape just like a pine-tree; not, of course, like one of our branching Scotch firs here, but like an Italian stone pine, with a long straight stem and a flat parasol-shaped top. Sometimes it was blackish, sometimes spotted; and the good Admiral Pliny, who was always curious about natural science, ordered his cutter and went away across the bay to see what it could be. Earthquake shocks had been very common for the last few days; but I do not suppose that Pliny had any notion that the earthquakes and the cloud had aught to do with each other. However, he soon found out that they had, and to his cost. When he got near the opposite shore some of the sailors met him and entreated him to turn back. Cinders and pumice-stones were falling down from the sky, and flames breaking out of the mountain above. But Pliny would go on: he said that if people were in danger, it was his duty to help them; and that he must see this strange cloud, and note down the different shapes into which it changed. But the hot ashes fell faster and faster; the sea ebbed out suddenly, and left them nearly dry, and Pliny turned away to a place called Stabiæ, to the house of his friend Pomponianus, who was just going to escape in a boat. Brave Pliny told him not to be afraid, ordered his bath like a true Roman gentleman, and then went into dinner with a cheerful face. Flames came down from the mountain, nearer and nearer as the night drew on; but Pliny persuaded his friend that they were only fires in some villages from which the peasants had fled, and then went to bed and slept soundly. However, in the middle of the night they found the courtyard being fast filled with cinders, and, if they had not woke up the Admiral in time, he would never have been able to get out of the house. The earthquake shocks grew stronger and fiercer, till the house was ready to fall; and Pliny and his friend, and the sailors and the slaves, all fled into the open fields, amid a shower of stones and cinders, tying pillows over their heads to prevent their being beaten down. The day had come by this time, but not the dawn—for it was still pitch dark as night. They went down to their boats upon the shore; but the sea raged so horribly that there was no getting on board of them. Then Pliny grew tired, and made his men spread a sail for him, and lay down on it; but there came down upon them a rush of flames, and a horrible smell of sulphur, and all ran for their lives. Some of the slaves tried to help the Admiral upon his legs; but he sank down again overpowered with the brimstone fumes, and so was left behind. When they came back again, there he lay dead, but with his clothes in order and his face as quiet as if he had been only sleeping. And that was the end of a brave and learned man—a martyr to duty and to the love of science.

But what was going on in the meantime? Under clouds of ashes, cinders, mud, lava, three of those happy cities were buried at once—Herculaneum, Pompeii, Stabiæ. They were buried just as the people had fled from them, leaving the furniture and the earthenware, often even jewels and gold, behind, and here and there among them a human being who had not had time to escape from the dreadful deluge of dust. The ruins of Herculaneum and Pompeii have been dug into since; and the paintings, especially in Pompeii, are found upon the walls still fresh, preserved from the air by the ashes which have covered them in. When you are older you perhaps will go to Naples, and see in its famous museum the curiosities which have been dug out of the ruined cities; and you will walk, I suppose, along the streets of Pompeii and see the wheel-tracks in the pavement, along which carts and chariots rumbled 2000 years ago. Meanwhile, if you go nearer home, to the Crystal Palace and to the Pompeian Court, as it is called, you will see an exact model of one of these old buried houses, copied even to the very paintings on the wells, and judge for yourself, as far as a little boy can judge, what sort of life these thoughtless, luckless people lived 2000 years ago.

And what had become of Vesuvius, the treacherous mountain? Half or more than half of the side of the old crater had been blown away, and what was left, which is now called the Monte Somma, stands in a half circle round the new cone and new crater which is burning at this very day. True, after that eruption which killed Pliny, Vesuvius fell asleep again, and did not awake for 134 years, and then again for 269 years but it has been growing more and more restless as the ages have passed on, and now hardly a year passes without its sending out smoke and stones from its crater, and streams of lava from its sides.

And now, I suppose, you will want to know what a volcano is like, and what a cone, and a crater, and lava are?

What a volcano is like, it is easy enough to show you; for they are the most simply and beautifully shaped of all mountains, and they are alike all over the world, whether they be large or small. Almost every volcano in the world, I believe, is, or has been once, of the shape which you see in the drawing opposite; even those volcanoes in the Sandwich Islands, of which you have often heard, which are now great lakes of boiling fire upon flat downs, without any cone to them at all. They, I believe, are volcanoes which have fallen in ages ago: just as in Java a whole burning mountain fell in on the night of the 11th of August, in the year 1772. Then, after a short and terrible earthquake, a bright cloud suddenly covered the whole mountain. The people who dwelt around it tried to escape; but before the poor souls could get away the earth sunk beneath their feet, and the whole mountain fell in and was swallowed up with a noise as if great cannon were being fired. Forty villages and nearly 3000 people were destroyed, and where the mountain had been was only a plain of red-hot stones. In the same way, in the year 1698, the top of a mountain in Quito fell in in a single night, leaving only two immense peaks of rock behind, and pouring out great floods of mud mixed with dead fish; for there are underground lakes among those volcanoes which swarm with little fish which never see the light.

But most volcanoes as I say, are, or have been, the shape of the one which you see here. This is Cotopaxi, in Quito, more than 19,000 feet in height. All those sloping sides are made of cinders and ashes, braced together, I suppose, by bars of solid lava-stone inside, which prevent the whole from crumbling down. The upper part, you see, is white with snow, as far down as a line which is 15,000 feet above the sea; for the mountain is in the tropics, close to the equator, and the snow will not lie in that hot climate any lower down. But now and then the snow melts off and rushes down the mountain side in floods of water and of mud, and the cindery cone of Cotopaxi stands out black and dreadful against the clear blue sky, and then the people of that country know what is coming. The mountain is growing so hot inside that it melts off its snowy covering; and soon it will burst forth with smoke and steam, and red-hot stones and earthquakes, which will shake the ground, and roars that will be heard, it may be, hundreds of miles away.

And now for the words cone, crater, lava. If I can make you understand those words, you will see why volcanoes must be in general of the shape of Cotopaxi.

Cone, crater, lava: those words make up the alphabet of volcano learning. The cone is the outside of a huge chimney; the crater is the mouth of it. The lava is the ore which is being melted in the furnace below, that it may flow out over the surface of the old land, and make new land instead.

And where is the furnace itself? Who can tell that? Under the roots of the mountains, under the depths of the sea; down "the path which no fowl knoweth, and which the vulture's eye hath not seen: the lion's whelp hath not trodden it, nor the fierce lion passed by it. There He putteth forth His hand upon the rock; He overturneth the mountain by the roots; He cutteth out rivers among the rocks; and His eye seeth every precious thing"—while we, like little ants, run up and down outside the earth, scratching, like ants, a few feet down, and calling that a deep ravine; or peeping a few feet down into the crater of a volcano, unable to guess what precious things may lie below—below even the fire which blazes and roars up through the thin crust of the earth. For of the inside of this earth we know nothing whatsoever: we only know that it is, on an average, several times as heavy as solid rock; but how that can be, we know not.

So let us look at the chimney, and what comes out of it; for we can see very little more.

Why is a volcano like a cone?

For the same cause for which a molehill is like a cone, though a very rough one; and that the little heaps which the burrowing beetles make on the moor, or which the ant-lions in France make in the sand, are all something in the shape of a cone, with a hole like a crater in the middle. What the beetle and the ant-lion do on a very little scale, the steam inside the earth does on a great scale. When once it has forced a vent into the outside air, it tears out the rocks underground, grinds them small against each other, often into the finest dust, and blasts them out of the hole which it has made. Some of them fall back into the hole, and are shot out again: but most of them fall round the hole, most of them close to it, and fewer of them farther off, till they are piled up in a ring round it, just as the sand is piled up round a beetle's burrow. For days, and weeks, and months this goes on; even it may be for hundreds of years: till a great cone is formed round the steam vent, hundreds or thousands of feet in height, of dust and stones, and of cinders likewise. For recollect, that when the steam has blown away the cold earth and rock near the surface of the ground, it begins blowing out the hot rocks down below, red-hot, white-hot, and at last actually melted. But these, as they are hurled into the cool air above, become ashes, cinders, and blocks of stone again, making the hill on which they fall bigger and bigger continually. And thus does wise Madam How stand in no need of bricklayers, but makes her chimneys build themselves.

And why is the mouth of the chimney called a crater?

Crater, as you know, is Greek for a cup. And the mouth of these chimneys, when they have become choked and stopped working, are often just the shape of a cup, or (as the Germans call them) kessels, which means kettles, or caldrons. I have seen some of them as beautifully and exactly rounded as if a cunning engineer had planned them, and had them dug out with the spade. At first, of course, their sides and bottom are nothing but loose stones, cinders, slag, ashes, such as would be thrown out of a furnace. But Madam How, who, whenever she makes an ugly desolate place, always tries to cover over its ugliness, and set something green to grow over it, and make it pretty once more, does so often and often by her worn-out craters. I have seen them covered with short sweet turf, like so many chalk downs. I have seen them, too, filled with bushes, which held woodcocks and wild boars. Once I came on a beautiful round crater on the top of a mountain, which was filled at the bottom with a splendid crop of potatoes. Though Madam How had not put them there herself, she had at least taught the honest Germans to put them there. And often Madam How turns her worn-out craters into beautiful lakes. There are many such crater-lakes in Italy, as you will see if ever you go there; as you may see in English galleries painted by Wilson, a famous artist who died before you were born. You recollect Lord Macaulay's ballad, "The Battle of the Lake Regillus"? Then that Lake Regillus (if I recollect right) is one of these round crater lakes. Many such deep clear blue lakes have I seen in the Eifel, in Germany; and many a curious plant have I picked on their shores, where once the steam blasted, and the earthquake roared, and the ash-clouds rushed up high into the heaven, and buried all the land around in dust, which is now fertile soil. And long did I puzzle to find out why the water stood in some craters, while others, within a mile of them perhaps, were perfectly dry. That I never found out for myself. But learned men tell me that the ashes which fall back into the crater, if the bottom of it be wet from rain, will sometimes "set" (as it is called) into a hard cement; and so make the bottom of the great bowl waterproof, as if it were made of earthenware.

But what gives the craters this cup-shape at first?

Think—While the steam and stones are being blown out, the crater is an open funnel, with more or less upright walls inside. As the steam grows weaker, fewer and fewer stones fall outside, and more and more fall back again inside. At last they quite choke up the bottom of the great round hole. Perhaps, too, the lava or melted rock underneath cools and grows hard, and that chokes up the hole lower down. Then, down from the round edge of the crater the stones and cinders roll inward more and more. The rains wash them down, the wind blows them down. They roll to the middle, and meet each other, and stop. And so gradually the steep funnel becomes a round cup. You may prove for yourself that it must be so, if you will try. Do you not know that if you dig a round hole in the ground, and leave it to crumble in, it is sure to become cup-shaped at last, though at first its sides may have been quite upright, like those of a bucket? If you do not know, get a trowel and make your little experiment.

And now you ought to understand what "cone" and "crater" mean. And more, if you will think for yourself, you may guess what would come out of a volcano when it broke out "in an eruption," as it is usually called. First, clouds of steam and dust (what you would call smoke); then volleys of stones, some cool, some burning hot; and at the last, because it lies lowest of all, the melted rock itself, which is called lava.

And where would that come out? At the top of the chimney? At the top of the cone?

No. Madam How, as I told you, usually makes things make themselves. She has made the chimney of the furnace make itself; and next she will make the furnace-door make itself.

The melted lava rises in the crater—the funnel inside the cone—but it never gets to the top. It is so enormously heavy that the sides of the cone cannot bear its weight, and give way low down. And then, through ashes and cinders, the melted lava burrows out, twisting and twirling like an enormous fiery earth-worm, till it gets to the air outside, and runs off down the mountain in a stream of fire. And so you may see (as are to be seen on Vesuvius now) two eruptions at once—one of burning stones above, and one of melted lava below.

And what is lava?

That, I think, I must tell you another time. For when I speak of it I shall have to tell you more about Madam How, and her ways of making the ground on which you stand, than I can say just now. But if you want to know (as I dare say you do) what the eruption of a volcano is like, you may read what follows. I did not see it happen; for I never had the good fortune of seeing a mountain burning, though I have seen many and many a one which has been burnt—extinct volcanoes, as they are called.

The man who saw it—a very good friend of mine, and a very good man of science also—went last year to see an eruption on Vesuvius, not from the main crater, but from a small one which had risen up suddenly on the outside of it; and he gave me leave (when I told him that I was writing for children) to tell them what he saw.

This new cone, he said, was about 200 feet high, and perhaps 80 or 100 feet across at the top. And as he stood below it (it was not safe to go up it) smoke rolled up from its top, "rosy pink below," from the glare of the caldron, and above "faint greenish or blueish silver of indescribable beauty, from the light of the moon." But more—By good chance, the cone began to send out, not smoke only, but brilliant burning stones. "Each explosion," he says, "was like a vast girandole of rockets, with a noise (such as rockets would make) like the waves on a beach, or the wind blowing through shrouds. The mountain was trembling the whole time. So it went on for two hours and more; sometimes eight or ten explosions in a minute, and more than 1000 stones in each, some as large as two bricks end to end. The largest ones mostly fell back into the crater; but the smaller ones being thrown higher, and more acted on by the wind, fell in immense numbers on the leeward slope of the cone" (of course, making it bigger and bigger, as I have explained already to you), and of course, as they were intensely hot and bright, making the cone look as if it too was red-hot. But it was not so, he says, really. The colour of the stones was rather "golden, and they spotted the black cone over with their golden showers, the smaller ones stopping still, the bigger ones rolling down, and jumping along just like hares." "A wonderful pedestal," he says, "for the explosion which surmounted it." How high the stones flew up he could not tell. "There was generally one which went much higher than the rest, and pierced upwards towards the moon, who looked calmly down, mocking such vain attempts to reach her." The large stones, of course, did not rise so high; and some, he says, "only just appeared over the rim of the cone, above which they came floating leisurely up, to show their brilliant forms and intense white light for an instant, and then subside again."

Try and picture that to yourselves, remembering that this was only a little side eruption, of no more importance to the whole mountain than the fall of a slate off the roof is of importance to the whole house. And then think how mean and weak man's fireworks, and even man's heaviest artillery, are compared with the terrible beauty and terrible strength of Madam How's artillery underneath our feet.

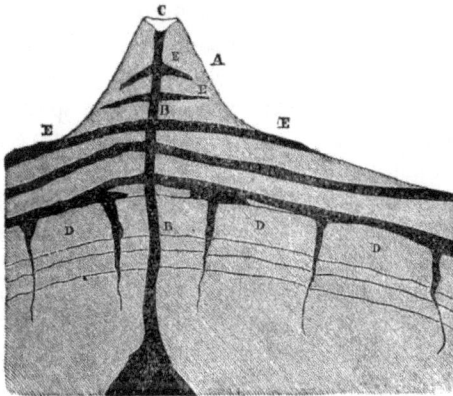

Now look at this figure. It represents a section of a volcano; that is, one cut in half to show you the inside. A is the cone of cinders. B, the black line up through the middle, is the funnel, or crack, through which steam, ashes, lava, and everything else rises. C is the crater mouth. D D

D, which looks broken, are the old rocks which the steam heaved up and burst before it could get out. And what are the black lines across, marked E E E? They are the streams of lava which have burrowed out, some covered up again in cinders, some lying bare in the open air, some still inside the cone, bracing it together, holding it up. Something like this is the inside of a volcano.

Chapter IV. The Transformations of a Grain of Soil

Why, you ask, are there such terrible things as volcanoes? Of what use can they be?

They are of use enough, my child; and of many more uses, doubt not, than we know as yet, or ever shall know. But of one of their uses I can tell you. They make, or help to make, divers and sundry curious things, from gunpowder to your body and mine.

What? I can understand their helping to make gunpowder, because the sulphur in it is often found round volcanoes; and I know the story of the brave Spaniard who, when his fellows wanted materials for gunpowder, had himself lowered in a basket down the crater of a South American volcano, and gathered sulphur for them off the burning cliffs: but how can volcanoes help to make me? Am I made of lava? Or is there lava in me?

My child, I did not say that volcanoes helped to make you. I said that they helped to make your body; which is a very different matter, as I beg you to remember, now and always. Your body is no more you yourself than the hoop which you trundle, or the pony which you ride. It is, like them, your servant, your tool, your instrument, your organ, with which you work: and a very useful, trusty, cunningly-contrived organ it is; and therefore I advise you to make good use of it, for you are responsible for it. But you yourself are not your body, or your brain, but something else, which we call your soul, your spirit, your life. And that "you yourself" would remain just the same if it were taken out of your body, and put into the body of a bee, or of a lion, or any other body; or into no body at all. At least so I believe; and so, I am happy to say, nine hundred and ninety-nine thousand nine hundred and ninety-nine people out of every million have always believed, because they have used their human instincts and their common sense, and have obeyed (without knowing it) the warning of a great and good philosopher called Herder, that "The organ is in no case the power which works by it;" which is as much as to say, that the engine is not the engine-driver, nor the spade the gardener.

There have always been, and always will be, a few people who cannot see that. They think that a man's soul is part of his body, and that he himself is not one thing, but a great number of things. They think that his mind and character are only made up of all the thoughts, and feelings, and recollections which have passed through his brain; and that as his brain changes, he himself must change, and become another person, and then another person again, continually. But do you not agree with them: but keep in mind wise Herder's warning that you are not to "confound the organ with the power," or the engine with the driver, or your body with yourself: and then we will go on and consider how a volcano, and the lava which flows from it, helps to make your body.

Now I know that the Scotch have a saying, "That you cannot make broth out of whinstones" (which is their name for lava). But, though they are very clever people, they are wrong there. I never saw any broth in Scotland, as far as I know, but what whinstones had gone to the making of it; nor a Scotch boy who had not eaten many a bit of whinstone, and been all the better for it.

Of course, if you simply put the whinstones into a kettle and boiled them, you would not get much out of them by such rough cookery as that. But Madam How is the best and most delicate of all cooks; and she knows how to pound, and soak, and stew whinstones so delicately, that she can make them sauce and seasoning for meat, vegetables, puddings, and almost everything that you eat; and can put into your veins things which were spouted up red-hot by volcanoes, ages and ages since, perhaps at the bottom of ancient seas which are now firm dry land.

This is very strange—as all Madam How's doings are. And you would think it stranger still if you had ever seen the flowing of a lava stream.

Out of a cave of slag and cinders in the black hillside rushes a golden river, flowing like honey, and yet so tough that you cannot thrust a stick into it, and so heavy that great stones (if you throw them on it) float on the top, and are carried down like corks on water. It is so hot that you cannot stand near it more than a few seconds; hotter, perhaps, than any fire you ever saw: but as it flows, the outside of it cools in the cool air, and gets covered with slag and cinders, something like those which you may see thrown out of the furnaces in the Black Country of Staffordshire. Sometimes these cling together above the lava stream, and make a tunnel, through the cracks in which you may see the fiery river rushing and roaring down below. But mostly they are kept broken and apart, and roll and slide over each other on the top of the lava, crashing and clanging as they grind together with a horrid noise. Of course that stream, like all streams, runs towards the lower grounds. It slides down glens, and fills them up; down the beds of streams, driving off the water in hissing steam; and sometimes (as it did in Iceland a few years ago) falls over some cliff, turning what had been a water-fall into a fire-fall, and filling up the pool below with blocks of lava suddenly cooled, with a clang and roar like that of chains shaken or brazen vessels beaten, which is heard miles and miles away. Of course, woe to the crops and gardens which stand in its way. It crawls over them all and eats them up. It shoves down houses; it sets woods on fire, and sends the steam and gas out of the tree-trunks hissing into the air. And (curiously enough) it does this often without touching the trees themselves. It flows round the trunks (it did so in a wood in the Sandwich Islands a few years ago), and of course sets them on fire by its heat, till nothing is left of them but blackened posts. But the moisture which comes out of the poor tree in steam blows so hard against the lava round that it can never touch the tree, and a round hole is left in the middle of the lava where the tree was. Sometimes, too, the lava will spit out liquid fire among the branches of the trees, which hangs down afterwards from them in tassels of slag, and yet, by the very same means, the steam in the branches will prevent the liquid fire burning them off, or doing anything but just scorch the bark.

But I can tell you a more curious story still. The lava stream, you must know, is continually sending out little jets of gas and steam: some of it it may have brought up from the very inside of the earth; most of it, I suspect, comes from the damp herbage and damp soil over which it runs. Be that as it may, a lava stream out of Mount Etna, in Sicily, came once down straight upon the town of Catania. Everybody thought that the town would be swallowed up; and the poor people there (who knew no better) began to pray to St. Agatha—a famous saint, who, they say, was martyred there ages ago—and who, they fancy, has power in heaven to save them from the lava stream. And really what happened was enough to make ignorant people, such as they were, think that St. Agatha had saved them. The lava stream came straight down upon the town wall. Another foot, and it would have touched it, and have begun shoving it down with a force compared with which all the battering-rams that you ever read of in ancient histories would be child's toys. But lo and behold! when the lava stream got within a few inches of the wall it stopped, and began to rear itself upright and build itself into a wall beside the wall. It rose and rose, till I believe in one place it overtopped the wall and began to curl over in a crest. All expected that it would fall over into the town at last: but no, there it stopped, and cooled, and hardened, and left the town unhurt. All the inhabitants said, of course, that St. Agatha had done it: but learned men found out that, as usual Madam How had done it, by making it do itself. The lava was so full of gas, which was continually blowing out in little jets, that when it reached the wall, it actually blew itself back from the wall; and, as the wall was luckily strong enough not to be blown down, the lava kept blowing itself back till it had time to cool. And so, my dear child, there was no miracle at all in the matter; and the poor people of Catania had to thank not St. Agatha, and any interference of hers, but simply Him who can preserve, just as He can destroy, by those laws of nature which are the breath of His mouth and the servants of His will.

But in many a case the lava does not stop. It rolls on and on over the downs and through the valleys, till it reaches the sea-shore, as it did in Hawaii in the Sandwich Islands this very year. And then it cools, of course; but often not before it has killed the fish by its sulphurous gases and heat, perhaps for miles around. And there is good reason to believe that the fossil fish which we so often find in rocks, perfect in every bone, lying sometimes in heaps, and twisted (as I have seen them) as if they had died suddenly and violently, were killed in this very way, either by heat from lava streams, or else by the bursting up of gases poisoning the water, in earthquakes and eruptions in the bottom of the sea. I could tell you many stories of fish being killed in thousands by earthquakes and volcanoes during the last few years. But we have not time to tell about everything.

And now you will ask me, with more astonishment than ever, what possible use can there be in these destroying streams of fire? And certainly, if you had ever seen a lava stream even when cool, and looked down, as I have done, at the great river of rough black blocks streaming away far and wide over the land, you would think it the most hideous and the most useless thing you ever saw. And yet, my dear child, there is One who told men to judge not according to the appearance, but to judge righteous judgment. He said that about matters spiritual and human: but it is quite as true about matters natural, which also are His work, and all obey His will.

Now if you had seen, as I have seen, close round the edges of these lava streams, and sometimes actually upon them, or upon the great bed of dust and ashes which have been hurled far and wide out of ancient volcanoes, happy homesteads, rich crops, hemp and flax, and wheat, tobacco, lucerne, roots, and vineyards laden with white and purple grapes, you would have begun to suspect that the lava streams were not, after all, such very bad neighbours. And when I tell you that volcanic soils (as they are called), that is, soil which has at first been lava or ashes, are generally the richest soils in the world—that, for instance (as some one told me the other day), there is soil in the beautiful island of Madeira so thin that you cannot dig more than two or three inches down without coming to the solid rock of lava, or what is harder even, obsidian (which is the black glass which volcanoes sometimes make, and which the old Mexicans used to chip into swords and arrows, because they had no steel)—and that this soil, thin as it is, is yet so fertile, that in it used to be grown the grapes of which the famous Madeira wine was made—when you remember this, and when you remember, too, the Lothians of Scotland (about which I shall have to say a little to you just now), then you will perhaps agree with me, that Lady Why has not been so very wrong in setting Madam How to pour out lava and ashes upon the surface of the earth.

For see—down below, under the roots of the mountains, Madam How works continually like a chemist in his laboratory, melting together all the rocks, which are the bones and leavings of the old worlds. If they stayed down below there, they would be of no use; while they will be of use up here in the open air. For, year by year—by the washing of rain and rivers, and also, I am sorry to say, by the ignorant and foolish waste of mankind—thousands and millions of tons of good stuff are running into the sea every year, which would, if it could be kept on land, make food for men and animals, plants and trees. So, in order to supply the continual waste of this upper world, Madam How is continually melting up the under world, and pouring it out of the volcanoes like manure, to renew the face of the earth. In these lava rocks and ashes which she sends up there are certain substances, without which men cannot live—without which a stalk of corn or grass cannot grow. Without potash, without magnesia, both of which are in your veins and mine—without silicates (as they are called), which give flint to the stems of corn and of grass, and so make them stiff and hard, and able to stand upright—and very probably without the carbonic acid gas, which comes out of the volcanoes, and is taken up by the leaves of plants, and turned by Madam How's cookery into solid wood—without all these things, and I suspect without a great many more things which come out of volcanoes—I do not see how this beautiful green world could get on at all.

Of course, when the lava first cools on the surface of the ground it is hard enough, and therefore barren enough. But Madam How sets to work upon it at once, with that delicate little water-spade of hers, which we call rain, and with that alone, century after century, and age after age, she digs the lava stream down, atom by atom, and silts it over the country round in rich manure. So that if Madam How has been a rough and hasty workwoman in pumping her treasures up out of her mine with her great steam-pumps, she shows herself delicate and tender and kindly enough in giving them away afterwards.

Nay, even the fine dust which is sometimes blown out of volcanoes is useful to countries far away. So light it is, that it rises into the sky and is wafted by the wind across the seas. So, in the year 1783, ashes from the Skaptar Jokull, in Iceland, were carried over the north of Scotland, and even into Holland, hundreds of miles to the south.

So, again, when in the year 1812 the volcano of St. Vincent, in the West India Islands, poured out torrents of lava, after mighty earthquakes which shook all that part of the world, a strange thing happened (about which I have often heard from those who saw it) in the island of Barbados, several hundred miles away. For when the sun rose in the morning (it was a Sunday morning), the sky remained more dark than any night, and all the poor negroes crowded terrified out of their houses into the streets, fancying the end of the world was come. But a learned man who was there, finding that, though the sun was risen, it was still pitchy dark, opened his window, and found that it was stuck fast by something on the ledge outside, and, when he thrust it open, found the ledge covered deep in soft red dust; and he instantly said, like a wise man as he was, "The volcano of St. Vincent must have broken out, and these are the ashes from it." Then he ran down stairs and quieted the poor negroes, telling them not to be afraid, for the end of the world was not coming just yet. But still the dust went on falling till the whole island, I am told, was covered an inch thick; and the same thing happened in the other islands round. People thought—and they had reason to think from what had often happened elsewhere—that though the dust might hurt the crops for that year, it would make them richer in years to come, because it would act as manure upon the soil; and so it did after a few years; but it did terrible damage at the time, breaking off the boughs of trees and covering up the crops; and in St. Vincent itself whole estates were ruined. It was a frightful day, but I know well that behind that How there was a Why for its happening, and happening too, about that very time, which all who know the history of negro slavery in the West Indies can guess for themselves, and confess, I hope, that in this case, as in all others, when Lady Why seems most severe she is often most just and kind.

Ah! my dear child, that I could go on talking to you of this for hours and days! But I have time now only to teach you the alphabet of these matters—and, indeed, I know little more than the alphabet myself; but if the very letters of Madam How's book, and the mere A, B, AB, of it, which I am trying to teach you, are so wonderful and so beautiful, what must its sentences be and its chapters? And what must the whole book be like? But that last none can read save He who wrote it before the worlds were made.

But now I see you want to ask a question. Let us have it out. I would sooner answer one question of yours than tell you ten things without your asking.

Is there potash and magnesia and silicates in the soil here? And if there is, where did they come from? For there are no volcanoes in England.

Yes. There are such things in the soil; and little enough of them, as the farmers here know too well. For we here, in Windsor Forest, are on the very poorest and almost the newest soil in England; and when Madam How had used up all her good materials in making the rest of the island, she carted away her dry rubbish and shot it down here for us to make the best of; and I do not think that we and our forefathers have done so very ill with it. But where the rich part, or staple, of our soils came from first it would be very difficult to say, so often has Madam How made, and unmade, and re-made England, and sifted her materials afresh every time. But if you go to the Lowlands of Scotland, you may soon see where the staple of the soil came from there, and that I was right in saying that there were atoms of lava in every Scotch boy's broth. Not that there were ever (as far as I know) volcanoes in Scotland or in England. Madam How has more than one string to her bow, or two strings either; so when she pours out her lavas, she does not always pour them out in the open air. Sometimes she pours them out at the bottom of the sea, as she did in the north of Ireland and the south-west of Scotland, when she made the Giant's Causeway, and Fingal's Cave in Staffa too, at the bottom of the old chalk ocean, ages and ages since. Sometimes she squirts them out between the layers of rock, or into cracks which the earthquakes have made, in what are called trap dykes, of which there are plenty to be seen in Scotland, and in Wales likewise. And then she lifts the earth up from the bottom of the sea, and sets the rain to wash away all the soft rocks, till the hard lava stands out in great hills upon the surface of the ground. Then the rain begins eating away those lava-hills likewise, and manuring the earth with them; and wherever those lava-hills stand up, whether great or small, there is pretty sure to be rich land around them. If you look at the Geological Map of England and Ireland, and the red spots upon it, which will show you where those old lavas are, you will see how much of them there is in England, at the Lizard Point in Cornwall, and how much more in Scotland and the north of Ireland. In South Devon, in Shropshire—with its beautiful Wrekin, and Caradoc, and Lawley—in Wales, round Snowdon (where some of the soil is very rich), and, above all, in the Lowlands of Scotland, you see these red marks, showing the old lavas, which are always fertile, except the poor old granite, which is of little use save to cut into building stone, because it is too full of quartz—that is, flint.

Think of this the next time you go through Scotland in the railway, especially when you get near Edinburgh. As you run through the Lothians, with their noble crops of corn, and roots, and grasses—and their great homesteads, each with its engine chimney, which makes steam do the work of men—you will see rising out of the plain, hills of dark rock, sometimes in single knobs, like Berwick Law or Stirling Crag—sometimes in noble ranges, like Arthur's Seat, or the Sidlaws, or the Ochils. Think what these black bare lumps of whinstone are, and what they do. Remember they are mines—not gold mines, but something richer still—food mines, which Madam How thrust into the inside of the earth, ages and ages since, as molten lava rock, and then cooled them and lifted them up, and pared them away with her ice-plough and her rain-spade, and spread the stuff of them over the wide carses round, to make in that bleak northern climate, which once carried nothing but fir-trees and heather, a soil fit to feed a great people; to cultivate in them industry, and science, and valiant self-dependence and self-help; and to gather round the Heart of Midlothian and the Castle Rock of Edinburgh the stoutest and the ablest little nation which Lady Why has made since she made the Greeks who fought at Salamis.

Of those Greeks you have read, or ought to read, in Mr. Cox's *Tales of the Persian War*. Some day you will read of them in their own books, written in their grand old tongue. Remember that Lady Why made them, as she has made the Scotch, by first preparing a country for them, which would call out all their courage and their skill; and then by giving them the courage and the skill to make use of the land where she had put them.

And now think what a wonderful fairy tale you might write for yourself—and every word of it true—of the adventures of one atom of Potash or some other Salt, no bigger than a needle's point, in such a lava stream as I have been telling of. How it has run round and round, and will run round age after age, in an endless chain of change. How it began by being molten fire underground, how then it became part of a hard cold rock, lifted up into a cliff, beaten upon by rain and storm, and washed down into the soil of the plain, till, perhaps, the little atom of mineral met with the rootlet of some great tree, and was taken up into its sap in spring, through tiny veins, and hardened the next year into a piece of solid wood. And then how that tree was cut down, and its logs, it may be, burnt upon the hearth, till the little atom of mineral lay among the wood-ashes, and was shoveled out and thrown upon the field and washed into the soil again, and taken up by the roots of a clover plant, and became an atom of vegetable matter once more. And then how, perhaps, a rabbit came by, and ate the clover, and the grain of mineral became part of the rabbit; and then how a hawk killed that rabbit, and ate it, and so the grain became part of the hawk; and how the farmer shot the hawk, and it fell perchance into a stream, and was carried down into the sea; and when its body decayed, the little grain sank through the water, and was mingled with the mud at the bottom of the sea. But do its wanderings stop there? Not so, my child. Nothing upon this earth, as I told you once before, continues in one stay. That grain of mineral might stay at the bottom of the sea a thousand or ten thousand years, and yet the time would come when Madam How would set to work on it again. Slowly, perhaps, she would sink that mud so deep, and cover it up with so many fresh beds of mud, or sand, or lime, that under the heavy weight, and perhaps, too, under the heat of the inside of the earth, that Mud would slowly change to hard Slate Rock; and ages after, it may be, Madam How might melt that Slate Rock once more, and blast it out; and then through the mouth of a volcano the little grain of mineral might rise into the open air again to make fresh soil, as it had done thousands of years before. For Madam How can manufacture many different things out of the same materials. She may have so wrought with that grain of mineral, that she may have formed it into part of a precious stone, and men may dig it out of the rock, or pick it up in the river-bed, and polish it, and set it, and wear it. Think of that—that in the jewels which your mother or your sisters wear, or in your father's signet ring, there may be atoms which were part of a live plant, or a live animal, millions of years ago, and may be parts of a live plant or a live animal millions of years hence.

Think over again, and learn by heart, the links of this endless chain of change: Fire turned into Stone—Stone into Soil—Soil into Plant—Plant into Animal—Animal into Soil—Soil into Stone—Stone into Fire again—and then Fire into Stone again, and the old thing run round once more.

So it is, and so it must be. For all things which are born in Time must change in Time, and die in Time, till that Last Day of this our little earth, in which,

> "Like to the baseless fabric of a vision,
> The cloud-capped towers, the gorgeous palaces,
> The solemn temples, the great globe itself,
> Yea, all things which inherit, shall dissolve,
> And, like an unsubstantial pageant faded,
> Leave not a rack behind."

So all things change and die, and so your body too must change and die—but not yourself. Madam How made your body; and she must unmake it again, as she unmakes all her works in Time and Space; but you, child, your Soul, and Life, and Self, she did not make; and over you she has no power. For you were not, like your body, created in Time and Space; and you will endure though Time and Space should be no more: because you are the child of the Living God, who gives to each thing its own body, and can give you another body, even as seems good to Him.

Chapter V. The Ice-Plough

You want to know why I am so fond of that little bit of limestone, no bigger than my hand, which lies upon the shelf; why I ponder over it so often, and show it to all sensible people who come to see me?

I do so, not only for the sake of the person who gave it to me, but because there is written on it a letter out of Madam How's alphabet, which has taken wise men many a year to decipher. I could not decipher that letter when first I saw the stone. More shame for me, for I had seen it often before, and understood it well enough, in many another page of Madam How's great book. Take the stone, and see if you can find out anything strange about it.

Well, it is only a bit of marble as big as my hand, that looks as if it had been, and really has been, broken off by a hammer. But when you look again, you see there is a smooth scraped part on one edge, that seems to have been rubbed against a stone.

Now look at that rubbed part, and tell me how it was done.

You have seen men often polish one stone on another, or scour floors with a Bath brick, and you will guess at first that this was polished so: but if it had been, then the rubbed place would have been flat: but if you put your fingers over it, you will find that it is not flat. It is rolled, fluted, channeled, so that the thing or things which rubbed it must have been somewhat round. And it is covered, too, with very fine and smooth scratches or grooves, all running over the whole in the same line. Now what could have done that?

Of course a man could have done it, if he had taken a large round stone in his hand, and worked the large channelings with that, and then had taken fine sand and gravel upon the points of his fingers, and worked the small scratches with that. But this stone came from a place where man had, perhaps, never stood before,—ay, which, perhaps, had never seen the light of day before since the world was made; and as I happen to know that no man made the marks upon that stone, we must set to work and think again for some tool of Madam How's which may have made them.

And now I think you must give up guessing, and I must tell you the answer to the riddle. Those marks were made by a hand which is strong and yet gentle, tough and yet yielding, like the hand of a man; a hand which handles and uses in a grip stronger than a giant's its own carving tools, from the great boulder stone as large as this whole room to the finest grain of sand. And that is ICE.

That piece of stone came from the side of the Rosenlaüi glacier in Switzerland, and it was polished by the glacier ice. The glacier melted and shrank this last hot summer farther back than it had done for many years, and left bare sheets of rock, which it had been scraping at for ages, with all the marks fresh upon them. And that bit was broken off and brought to me, who never saw a glacier myself, to show me how the marks which the ice makes in Switzerland are exactly the same as those which the ice has made in Snowdon and in the Highlands, and many another place where I have traced them, and written a little, too, about them in years gone by. And so I treasure this, as a sign that Madam How's ways do not change nor her laws become broken; that, as that great philosopher Sir Charles Lyell will tell you, when you read his books, Madam How is making and unmaking the surface of the earth now, by exactly the same means as she was making and unmaking ages and ages since; and that what is going on slowly and surely in the Alps in Switzerland was going on once here where we stand.

It is very difficult, I know, for a little boy like you to understand how ice, and much more how soft snow, should have such strength that it can grind this little stone, much more such strength as to grind whole mountains into plains. You have never seen ice and snow do harm. You cannot even recollect the Crimean Winter, as it was called then; and well for you you cannot, considering all the misery it brought at home and abroad. You cannot, I say, recollect the Crimean Winter, when the Thames was frozen over above the bridges, and the ice piled in little bergs ten to fifteen feet high, which lay, some of them, stranded on the shores, about London itself, and did not melt, if I recollect, until the end of May. You never stood, as I stood, in the great winter of 1837-8 on Battersea Bridge, to see the ice break up with the tide, and saw the great slabs and blocks leaping and piling upon each other's backs, and felt the bridge tremble with their shocks, and listened to their horrible grind and roar, till one got some little picture in one's mind of what must be the breaking up of an ice-floe in the Arctic regions, and what must be the danger of a ship nipped in the ice and lifted up on high, like those in the pictures of Arctic voyages which you are so fond of looking through. You cannot recollect how that winter even in our little Blackwater Brook the alder stems were all peeled white, and scarred, as if they had been gnawed by hares and deer, simply by the rushing and scraping of the ice,—a sight which gave me again a little picture of the destruction which the ice makes of quays, and stages, and houses along the shore upon the coasts of North America, when suddenly setting in with wind and tide, it jams and piles up high inland, as you may read for yourself some day in a delightful book called *Frost and Fire*. You recollect none of these things. Ice and snow are to you mere playthings; and you long for winter, that you may make snowballs and play hockey and skate upon the ponds, and eat ice like a foolish boy till you make your stomach ache. And I dare say you have said, like many another boy, on a bright cheery ringing frosty day, "Oh, that it would be always winter!" You little knew for what you asked. You little thought what the earth would soon be like, if it were always winter,—if one sheet of ice on the pond glued itself on to the bottom of the last sheet, till the whole pond was a solid mass,—if one snow-fall lay upon the top of another snow-fall till the moor was covered many feet deep and the snow began sliding slowly down the glen from Coombs's, burying the green fields, tearing the trees up by their roots, burying gradually house, church, and village, and making this place for a few thousand years what it was many thousand years ago. Good-bye then, after a very few winters, to bees, and butterflies, and singing-birds, and flowers; and good-bye to all vegetables, and fruit, and bread; good-bye to cotton and woollen clothes. You would have, if you were left alive, to dress in skins, and eat fish and seals, if any came near enough to be caught. You would have to live in a word, if you could live at all, as Esquimaux live now in Arctic regions, and as people had to live in England ages since, in the times when it was always winter, and icebergs floated between here and Finchampstead. Oh no, my child: thank Heaven that it is not always winter; and remember that winter ice and snow, though it is a very good tool with which to make the land, must leave the land year by year if that land is to be fit to live in.

I said that if the snow piled high enough upon the moor, it would come down the glen in a few years through Coombs's Wood; and I said then you would have a small glacier here—such a glacier (to compare small things with great) as now comes down so many valleys in the Alps, or has come down all the valleys of Greenland and Spitzbergen till they reach the sea, and there end as cliffs of ice, from which great icebergs snap off continually, and fall and float away, wandering southward into the Atlantic for many a hundred miles. You have seen drawings of such glaciers in Captain Cook's Voyages; and you may see photographs of Swiss glaciers in any good London print-shop; and therefore you have seen almost as much about them as I have seen, and may judge for yourself how you would like to live where it is always winter.

Now you must not ask me to tell you what a glacier is like, for I have never seen one; at least, those which I have seen were more than fifty miles away, looking like white clouds hanging on the gray mountain sides. And it would be an impertinence—that means a meddling with things which I have no business—to picture to you glaciers which have been pictured so well and often by gentlemen who escape every year from their hard work in town to find among the glaciers of the Alps health and refreshment, and sound knowledge, and that most wholesome and strengthening of all medicines, toil.

So you must read of them in such books as *Peaks, Passes, and Glaciers*, and Mr. Willes's *Wanderings in the High Alps*, and Professor Tyndall's different works; or you must look at them (as I just now said) in photographs or in pictures. But when you do that, or when you see a glacier for yourself, you must bear in mind what a glacier means—that it is a river of ice, fed by a lake of snow. The lake from which it springs is the eternal snow-field which stretches for miles and miles along the mountain tops, fed continually by fresh snow-storms falling from the sky. That snow slides off into the valleys hour by hour, and as it rushes down is ground and pounded, and thawed and frozen again into a sticky paste of ice, which flows slowly but surely till it reaches the warm valley at the mountain foot, and there melts bit by bit. The long black lines which you see winding along the white and green ice of the glacier are the stones which have fallen from the cliffs above. They will be dropped at the end of the glacier, and mixed with silt and sand and other stones which have come down inside the glacier itself, and piled up in the field in great mounds, which are called moraines, such as you may see and walk on in Scotland many a time, though you might never guess what they are.

The river which runs out at the glacier foot is, you must remember, all foul and milky with the finest mud; and that mud is the grinding of the rocks over which the glacier has been crawling down, and scraping them as it scraped my bit of stone with pebbles and with sand. And this is the alphabet, which, if you learn by heart, you will learn to understand how Madam How uses her great ice-plough to plough down her old mountains, and spread the stuff of them about the valleys to make rich straths of fertile soil. Nay, so immensely strong, because immensely heavy, is the share of this her great ice-plough, that some will tell you (and it is not for me to say that they are wrong) that with it she has ploughed out all the mountain lakes in Europe and in North America; that such lakes, for instance, as Ullswater or Windermere have been scooped clean out of the solid rock by ice which came down these glaciers in old times. And be sure of this, that next to Madam How's steam-pump and her rain-spade, her great ice-plough has had, and has still, the most to do with making the ground on which we live.

Do I mean that there were ever glaciers here? No, I do not. There have been glaciers in Scotland in plenty. And if any Scotch boy shall read this book, it will tell him presently how to find the marks of them far and wide over his native land. But as you, my child, care most about this country in which you live, I will show you in any gravel-pit, or hollow lane upon the moor, the marks, not of a glacier, which is an ice-river, but of a whole sea of ice.

Let us come up to the pit upon the top of the hill, and look carefully at what we see there. The lower part of the pit of course is a solid rock of sand. On the top of that is a cap of gravel, five, six, ten feet thick. Now the sand was laid down there by water at the bottom of an old sea; and therefore the top of it would naturally be flat and smooth, as the sands at Hunstanton or at Bournemouth are; and the gravel, if it was laid down by water, would naturally lie flat on it again: but it does not. See how the top of the sand is dug out into deep waves and pits, filled up with gravel. And see, too, how over some of the gravel you get sand again, and then gravel again, and then sand again, till you cannot tell where one fairly begins and the other ends. Why, here are little dots of gravel, six or eight feet down, in what looks the solid sand rock, yet the sand must have been opened somehow to put the gravel in.

You say you have seen that before. You have seen the same curious twisting of the gravel and sand into each other on the top of Farley Hill, and in the new cutting on Minley Hill; and, best of all, in the railway cutting between Ascot and Sunningdale, where upon the top the white sand and gravel is arranged in red and brown waves, and festoons, and curlicues, almost like Prince of Wales's feathers. Yes, that last is a beautiful section of ice-work; so beautiful, that I hope to have it photographed some day.

Now, how did ice do this?

Well, I was many a year before I found out that, and I dare say I never should have found it out for myself. A gentleman named Trimmer, who, alas! is now dead, was, I believe, the first to find it out.

He knew that along the coast of Labrador, and other cold parts of North America, and on the shores, too, of the great river St. Lawrence, the stranded icebergs, and the ice-foot, as it is called, which is continually forming along the freezing shores, grub and plough every tide into the mud and sand, and shove up before them, like a ploughshare, heaps of dirt; and that, too, the ice itself is full of dirt, of sand and stones, which it may have brought from hundreds of miles away; and that, as this ploughshare of dirty ice grubs onward, the nose of the plough is continually being broken off, and left underneath the mud; and that, when summer comes, and the ice melts, the mud falls back into the place where the ice had been, and covers up the gravel which was in the ice. So, what between the grubbing of the ice-plough into the mud, and the dirt which it leaves behind when it melts, the stones, and sand, and mud upon the shore are jumbled up into curious curved and twisted layers, exactly like those which Mr. Trimmer saw in certain gravel-pits. And when I first read about that, I said, "And exactly like what I have been seeing in every gravel-pit round here, and trying to guess how they could have been made by currents of water, and yet never could make any guess which would do." But after that it was all explained to me; and I said, "Honour to the man who has let Madam How teach him what she had been trying to teach me for fifteen years, while I was too stupid to learn it. Now I am certain, as certain as I can be of any earthly thing, that the whole of these Windsor Forest Flats were ages ago ploughed and harrowed over and over again, by ice-floes and icebergs drifting and stranding in a shallow sea."

And if you say, my dear child, as some people will say, that it is like building a large house upon a single brick to be sure that there was an iceberg sea here, just because I see a few curlicues in the gravel and sand—then I must tell you that there are sometimes—not often, but sometimes—pages in Madam How's book in which one single letter tells you as much as a whole chapter; in which if you find one little fact, and know what it really means, it makes you certain that a thousand other great facts have happened. You may be astonished: but you cannot deny your own eyes, and your own common sense. You feel like Robinson Crusoe when, walking along the shore of his desert island, he saw for the first time the print of a man's foot in the sand. How it could have got there without a miracle he could not dream. But there it was. One footprint was as good as the footprints of a whole army would have been. A man had been there; and more men might come. And in fear of the savages—and if you have read Robinson Crusoe you know how just his fears were—he went home trembling and loaded his muskets, and barricaded his cave, and passed sleepless nights watching for the savages who might come, and who came after all.

And so there are certain footprints in geology which there is no mistaking; and the prints of the ice-plough are among them.

For instance:—When they were trenching the new plantation close to Wellington College station, the men turned up out of the ground a great many Sarsden stones; that is, pieces of hard sugary sand, such as Stonehenge is made of. And when I saw these I said, "I suspect these were brought here by icebergs:" but I was not sure, and waited. As the men dug on, they dug up a great many large flints, with bottle-green coats. "Now," I said, "I am sure. For I know where these flints must have come from." And for reasons which would be too long to tell you here, I said, "Some time or other, icebergs have been floating northward from the Hog's Back over Aldershot and Farnborough, and have been trying to get into the Vale of Thames by the slope at Wellington College station; and they have stranded, and dropped these flints." And I am so sure of that, that if I found myself out wrong after all I should be at my wit's end; for I should know that I was wrong about a hundred things besides.

Or again, if you ever go up Deeside in Scotland, towards Balmoral, and turn up Glen Muick, towards Alt-na-guisach, of which you may see a picture in the Queen's last book, you will observe standing on your right hand, just above Birk Hall, three pretty rounded knolls, which they call the Coile Hills. You may easily know them by their being covered with beautiful green grass instead of heather. That is because they are made of serpentine or volcanic rock, which (as you have seen) often cuts into beautiful red and green marble; and which also carries a very rich soil because it is full of magnesia. If you go up those hills, you get a glorious view—the mountains sweeping round you where you stand, up to the top of Lochnagar, with its bleak walls a thousand feet perpendicular, and gullies into which the sun never shines, and round to the dark fir forests of the Ballochbuie. That is the arc of the bow; and the cord of the bow is the silver Dee, more than a thousand feet below you; and in the centre of the cord, where the arrow would be fitted in, stands Balmoral, with its Castle, and its Gardens, and its Park, and pleasant cottages and homesteads all around. And when you have looked at the beautiful amphitheatre of forest at your feet, and looked too at the great mountains to the westward, and Benaun, and Benna-buird and Benna-muicdhui, with their bright patches of eternal snow, I should advise you to look at the rock on which you stand, and see what you see there. And you will see that on the side of the Coiles towards Lochnagar, and between the knolls of them, are scattered streams, as it were, of great round boulder stones—which are not serpentine, but granite from the top of Lochnagar, five miles away. And you will see that the knolls of serpentine rock, or at least their backs and shoulders towards Lochnagar, are all smoothed and polished till they are as round as the backs of sheep, "roches moutonnées," as the French call ice-polished rocks; and then, if you understand what that means, you will say, as I said, "I am perfectly certain that this great basin between me and Lochnagar, which is now 3000 feet deep of empty air was once filled up with ice to the height of the hills on which I stand—about 1700 feet high—and that that ice ran over into Glen Muick, between these pretty knolls, and covered the ground where Birk Hall now stands."

And more:—When you see growing on those knolls of serpentine a few pretty little Alpine plants, which have no business down there so low, you will have a fair right to say, as I said, "The seeds of these plants were brought by the ice ages and ages since from off the mountain range of Lochnagar, and left here, nestling among the rocks, to found a fresh colony, far from their old mountain home."

If I could take you with me up to Scotland,—take you, for instance, along the Tay, up the pass of Dunkeld, or up Strathmore towards Aberdeen, or up the Dee towards Braemar,—I could show you signs, which cannot be mistaken, of the time when Scotland was, just like Spitzbergen or like Greenland now, covered in one vast sheet of snow and ice from year's end to year's end; when glaciers were ploughing out its valleys, icebergs were breaking off the icy cliffs and floating out to sea; when not a bird, perhaps, was to be seen save sea-fowl, not a plant upon the rocks but a few lichens, and Alpine saxifrages, and such like—desolation and cold and lifeless everywhere. That ice-time went on for ages and for ages; and yet it did not go on in vain. Through it Madam How was ploughing down the mountains of Scotland to make all those rich farms which stretch from the north side of the Frith of Forth into Sutherlandshire. I could show you everywhere the green banks and knolls of earth, which Scotch people call "kames" and "tomans"—perhaps brought down by ancient glaciers, or dropped by ancient icebergs—now so smooth and green through summer and through winter, among the wild heath and the rough peat-moss, that the old Scots fancied, and I dare say Scotch children fancy still, fairies dwelt inside. If you laid your ear against the mounds, you might hear the fairy music, sweet and faint, beneath the ground. If you watched the mound at night, you might see the fairies dancing the turf short and smooth, or riding out on fairy horses, with green silk clothes and jingling bells. But if you fell asleep upon the mounds, the fairy queen came out and carried you for seven years into Fairyland, till you awoke again in the same place, to find all changed around you, and yourself grown thin and old.

These are all dreams and fancies—untrue, not because they are too strange and wonderful, but because they are not strange and wonderful enough: for more wonderful sure than any fairy tale it is, that Madam How should make a rich and pleasant land by the brute force of ice.

And were there any men and women in that old age of ice? That is a long story, and a dark one too; we will talk of it next time.

Chapter VI. The True Fairy Tale

You asked if there were men in England when the country was covered with ice and snow. Look at this, and judge for yourself.

What is it? a piece of old mortar? Yes. But mortar which was made Madam How herself, and not by any man. And what is in it? A piece of flint and some bits of bone. But look at that piece of flint. It is narrow, thin, sharp-edged: quite different in shape from any bit of flint which you or I ever saw among the hundreds of thousands of broken bits of gravel which we tread on here all day long; and here are some more bits like it, which came from the same place—all very much the same shape, like rough knives or razor blades; and here is a core of flint, the remaining part of a large flint, from which, as you may see, blades like those have been split off. Those flakes of flint, my child, were split off by men; even your young eyes ought to be able to see that. And here are other pieces of flint—pear-shaped, but flattened, sharp at one end and left rounded at the other, which look like spear-heads, or arrow-heads, or pointed axes, or pointed hatchets—even your young eyes can see that these must have been made by man. And they are, I may tell you, just like the tools of flint, or of obsidian, which is volcanic glass, and which savages use still where they have not iron. There is a great obsidian knife, you know, in a house in this very parish, which came from Mexico; and your eye can tell you how like it is to these flint ones. But these flint tools are very old. If you crack a fresh flint, you will see that its surface is gray, and somewhat rough, so that it sticks to your tongue. These tools are smooth and shiny: and the edges of some of them are a little rubbed from being washed about in gravel; while the iron in the gravel has stained them reddish, which it would take hundreds and perhaps thousands of years to do. There are little rough markings, too, upon some of them, which, if you look at through a

magnifying glass, are iron, crystallised into the shape of little sea-
weeds and trees—another sign that they are very very old. And what is
more, near the place where these flint flakes come from there are no
flints in the ground for hundreds of miles; so that men must have
brought them there ages and ages since. And to tell you plainly, these
are scrapers such as the Esquimaux in North America still use to scrape
the flesh off bones, and to clean the insides of skins.

But did these people (savages perhaps) live when the country was
icy cold? Look at the bits of bone. They have been split, you see,
lengthways; that, I suppose, was to suck the marrow out of them, as
savages do still. But to what animal do the bones belong? That is the
question, and one which I could not have answered you, if wiser men
than I am could not have told me.

They are the bones of reindeer—such reindeer as are now found
only in Lapland and the half-frozen parts of North America, close to the
Arctic circle, where they have six months day and six months night.
You have read of Laplanders, and how they drive reindeer in their
sledges, and live upon reindeer milk; and you have read of Esquimaux,
who hunt seals and walrus, and live in houses of ice, lighted by lamps
fed with the same blubber on which they feed themselves. I need not
tell you about them.

Now comes the question—Whence did these flints and bones
come? They came out of a cave in Dordogne, in the heart of sunny
France,—far away to the south, where it is hotter every summer than it
was here even this summer, from among woods of box and evergreen
oak, and vineyards of rich red wine. In that warm land once lived

savages, who hunted amid ice and snow the reindeer, and with the reindeer animals stranger still.

And now I will tell you a fairy tale: to make you understand it at all I must put it in the shape of a tale. I call it a fairy tale, because it is so strange; indeed I think I ought to call it the fairy tale of all fairy tales, for by the time we get to the end of it I think it will explain to you how our forefathers got to believe in fairies, and trolls, and elves, and scratlings, and all strange little people who were said to haunt the mountains and the caves.

Well, once upon a time, so long ago that no man can tell when, the land was so much higher, that between England and Ireland, and, what is more, between England and Norway, was firm dry land. The country then must have looked—at least we know it looked so in Norfolk— very like what our moors look like here. There were forests of Scotch fir, and of spruce too, which is not wild in England now, though you may see plenty in every plantation. There were oaks and alders, yews and sloes, just as there are in our woods now. There was buck-bean in the bogs, as there is in Larmer's and Heath pond; and white and yellow water-lilies, horn-wort, and pond-weeds, just as there are now in our ponds. There were wild horses, wild deer, and wild oxen, those last of an enormous size. There were little yellow roe-deer, which will not surprise you, for there are hundreds and thousands in Scotland to this day; and, as you know, they will thrive well enough in our woods now. There were beavers too: but that must not surprise you, for there were beavers in South Wales long after the Norman Conquest, and there are beavers still in the mountain glens of the south-east of France. There were honest little water-rats too, who I dare say sat up on their hind legs like monkeys, nibbling the water-lily pods, thousands of years ago, as they do in our ponds now. Well, so far we have come to nothing strange: but now begins the fairy tale. Mixed with all these animals, there wandered about great herds of elephants and rhinoceroses; not smooth-skinned, mind, but covered with hair and wool, like those which are still found sticking out of the everlasting ice cliffs, at the mouth of the Lena and other Siberian rivers, with the flesh, and skin, and hair so fresh upon them, that the wild wolves tear it off, and snarl and growl over the carcass of monsters who were frozen up thousands of years ago. And with them, stranger still, were great hippopotamuses; who came, perhaps, northward in summer time along the sea-shore and down the rivers, having spread hither all the way from Africa; for in those days, you must understand, Sicily, and Italy, and Malta—look at your map—were joined to the coast of Africa: and so it may be was the rock of Gibraltar itself; and over the sea where the Straits of Gibraltar now flow was firm dry land, over which hyenas and leopards, elephants and rhinoceroses ranged into Spain; for their bones are found at this day in the Gibraltar caves. And this is the first chapter of my fairy tale.

Now while all this was going on, and perhaps before this began, the climate was getting colder year by year—we do not know how; and, what is more, the land was sinking; and it sank so deep that at last nothing was left out of the water but the tops of the mountains in Ireland, and Scotland, and Wales. It sank so deep that it left beds of shells belonging to the Arctic regions nearly two thousand feet high upon the mountain side. And so

> "It grew wondrous cold,
> And ice mast-high came floating by,
> As green as emerald."

But there were no masts then to measure the icebergs by, nor any ship nor human being there. All we know is that the icebergs brought with them vast quantities of mud, which sank to the bottom, and covered up that pleasant old forest-land in what is called boulder-clay; clay full of bits of broken rock, and of blocks of stone so enormous, that nothing but an iceberg could have carried them. So all the animals were drowned or driven away, and nothing was left alive perhaps, except a few little hardy plants which clung about cracks and gullies in the mountain tops; and whose descendants live there still. That was a dreadful time; the worst, perhaps, of all the age of Ice; and so ends the second chapter of my fairy tale.

Now for my third chapter. "When things come to the worst," says the proverb, "they commonly mend;" and so did this poor frozen and drowned land of England and France and Germany, though it mended very slowly. The land began to rise out of the sea once more, and rose till it was perhaps as high as it had been at first, and hundreds of feet higher than it is now: but still it was very cold, covered, in Scotland at least, with one great sea of ice and glaciers descending down into the sea, as I said when I spoke to you about the Ice-Plough. But as the land rose, and grew warmer too, while it rose, the wild beasts who had been driven out by the great drowning came gradually back again. As the bottom of the old icy sea turned into dry land, and got covered with grasses, and weeds, and shrubs once more, elephants, rhinoceroses, hippopotamuses, oxen—sometimes the same species, sometimes slightly different ones—returned to France, and then to England (for there was no British Channel then to stop them); and with them came other strange animals, especially the great Irish elk, as he is called, as large as the largest horse, with horns sometimes ten feet across. A pair of those horns with the skull you have seen yourself, and can judge what a noble animal he must have been. Enormous bears came too, and hyænas, and a tiger or lion (I cannot say which), as large as the largest Bengal tiger now to be seen in India.

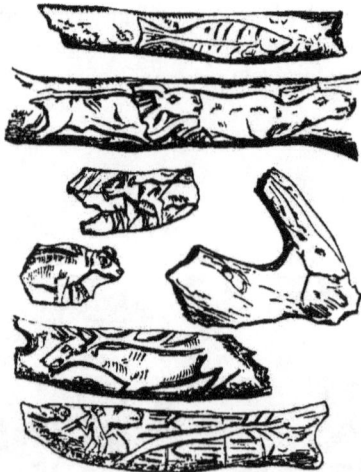

And in those days—we cannot, of course, exactly say when—there came—first I suppose into the south and east of France, and then gradually onward into England and Scotland and Ireland—creatures without any hair to keep them warm, or scales to defend them, without horns or tusks to fight with, or teeth to worry and bite; the weakest you would have thought of the beasts, and yet stronger than all the animals, because they were Men, with reasonable souls. Whence they came we cannot tell, nor why; perhaps from mere hunting after food, and love of wandering and being independent and alone. Perhaps they came into that icy land for fear of stronger and cleverer people than themselves; for we have no proof, my child, none at all, that they were the first men that trod this earth. But be that as it may, they came; and so cunning were these savage men, and so brave likewise, though they had no iron among them, only flint and sharpened bones, yet they contrived to kill and eat the mammoths, and the giant oxen, and the wild horses, and the reindeer, and to hold their own against the hyenas, and tigers, and bears, simply because they had wits, and the dumb animals had none. And that is the strangest part to me of all my fairy tale. For what a man's wits are, and why he has them, and therefore is able to invent and to improve, while even the cleverest ape has none, and therefore can invent and improve nothing, and therefore cannot better himself, but must remain from father to son, and father to son again, a stupid, pitiful, ridiculous ape, while men can go on civilizing themselves, and growing richer and more comfortable, wiser and happier, year by year—how that comes to pass, I say, is to me a wonder and a prodigy and a miracle, stranger than all the most fantastic marvels you ever read in fairy tales.

You may find the flint weapons which these old savages used buried in many a gravel-pit up and down France and the south of England; but you will find none here, for the gravel here was made (I am told) at the beginning of the ice-time, before the north of England sunk into the sea, and therefore long, long before men came into this land. But most of their remains are found in caves which water has eaten out of the limestone rocks, like that famous cave of Kent's Hole at Torquay. In it, and in many another cave, lie the bones of animals which the savages ate, and cracked to get the marrow out of them, mixed up with their flint-weapons and bone harpoons, and sometimes with burnt ashes and with round stones, used perhaps to heat water, as savages do now, all baked together into a hard paste or breccia by the lime. These are in the water, and are often covered with a floor of stalagmite which has dripped from the roof above and hardened into stone. Of these caves and their beautiful wonders I must tell you another day. We must keep now to our fairy tale. But in these caves, no doubt, the savages lived; for not only have weapons been found in them, but actually drawings scratched (I suppose with flint) on bone or mammoth ivory—drawings of elk, and bull, and horse, and ibex—and one, which was found in France, of the great mammoth himself, the woolly elephant, with a mane on his shoulders like a lion's mane. So you see that one of the earliest fancies of this strange creature, called man, was to draw, as you and your schoolfellows love to draw, and copy what you see, you know not why. Remember that. You like to draw; but why you like it neither you nor any man can tell. It is one of the mysteries of human nature; and that poor savage clothed in skins, dirty it may be, and more ignorant than you (happily) can conceive, when he sat scratching on ivory in the cave the figures of the animals he hunted, was proving thereby that he had the same wonderful and mysterious human nature as you—that he was the kinsman of every painter and sculptor who ever felt it a delight and duty to copy the beautiful works of God.

Sometimes, again, especially in Denmark, these savages have left behind upon the shore mounds of dirt, which are called there "kjökken-möddings"—"kitchen-middens" as they would say in Scotland, "kitchen-dirtheaps" as we should say here down South—and a very good name for them that is; for they are made up of the shells of oysters, cockles, mussels, and periwinkles, and other shore-shells besides, on which those poor creatures fed; and mingled with them are broken bones of beasts, and fishes, and birds, and flint knives, and axes, and sling stones; and here and there hearths, on which they have cooked their meals in some rough way. And that is nearly all we know about them; but this we know from the size of certain of the shells, and from other reasons which you would not understand, that these mounds were made an enormous time ago, when the water of the Baltic Sea was far more salt than it is now.

But what has all this to do with my fairy tale? This:—

Suppose that these people, after all, had been fairies?

I am in earnest. Of course, I do not mean that these folk could make themselves invisible, or that they had any supernatural powers—any more, at least, than you and I have—or that they were anything but savages; but this I do think, that out of old stories of these savages grew up the stories of fairies, elves, and trolls, and scratlings, and cluricaunes, and ogres, of which you have read so many.

When stronger and bolder people, like the Irish, and the Highlanders of Scotland, and the Gauls of France, came northward with their bronze and iron weapons; and still more, when our own forefathers, the Germans and the Norsemen, came, these poor little savages with their flint arrows and axes, were no match for them, and had to run away northward, or to be all killed out; for people were fierce and cruel in those old times, and looked on every one of a different race from themselves as a natural enemy. They had not learnt—alas! too many have not learned it yet—that all men are brothers for the sake of Jesus Christ our Lord. So these poor savages were driven out, till none were left, save the little Lapps up in the north of Norway, where they live to this day.

But stories of them, and of how they dwelt in caves, and had strange customs, and used poisoned weapons, and how the elf-bolts (as their flint arrow-heads are still called) belonged to them, lingered on, and were told round the fire on winter nights and added to, and played with half in fun, till a hundred legends sprang up about them, which used once to be believed by grown-up folk, but which now only amuse children. And because some of these savages were very short, as the Lapps and Esquimaux are now, the story grew of their being so small that they could make themselves invisible; and because others of them were (but probably only a few) very tall and terrible, the story grew that there were giants in that old world, like that famous Gogmagog, whom Brutus and his Britons met (so old fables tell), when they landed first at Plymouth, and fought him, and threw him over the cliff. Ogres, too—of whom you read in fairy tales—I am afraid that there were such people once, even here in Europe; strong and terrible savages, who ate human beings. Of course, the legends and tales about them became ridiculous and exaggerated as they passed from mouth to mouth over the Christmas fire, in the days when no one could read or write. But that the tales began by being true any one may well believe who knows how many cannibal savages there are in the world even now. I think that, if ever there was an ogre in the world, he must have been very like a certain person who lived, or was buried, in a cave in the Neanderthal, between Elberfeld and Dusseldorf, on the Lower Rhine. The skull and bones which were found there (and which are very famous now among scientific men) belonged to a personage whom I should have been very sorry to meet, and still more to let you meet, in the wild forest; to a savage of enormous strength of limb (and I suppose of jaw) likewise

> "like an ape,
> With forehead villainous low,"

who could have eaten you if he would; and (I fear) also would have eaten you if he could. Such savages may have lingered (I believe, from the old ballads and romances, that they did linger) for a long time in lonely forests and mountain caves, till they were all killed out by warriors who wore mail-armour and carried steel sword, and battle-axe, and lance.

But had these people any religion?

My dear child, we cannot know, and need not know. But we know this—that God beholds all the heathen. He fashions the hearts of them, and understandeth all their works. And we know also that He is just and good. These poor folks were, I doubt not, happy enough in their way; and we are bound to believe (for we have no proof against it), that most of them were honest and harmless enough likewise. Of course, ogres and cannibals, and cruel and brutal persons (if there were any among them), deserved punishment—and punishment, I do not doubt, they got. But, of course, again, none of them knew things which you know; but for that very reason they were not bound to do many things which you are bound to do. For those to whom little is given, of them shall little be required. What their religion was like, or whether they had any religion at all, we cannot tell. But this we can tell, that known unto God are all His works from the creation of the world; and that His mercy is over all His works, and He hateth nothing that He has made. These men and women, whatever they were, were God's work; and therefore we may comfort ourselves with the certainty that, whether or not they knew God, God knew them.

And so ends my fairy tale.

But is it not a wonderful tale? More wonderful, if you will think over it, than any story invented by man. But so it always is. "Truth," wise men tell us, "is stranger than fiction." Even a child like you will see that it must be so, if you will but recollect who makes fiction, and who makes facts.

Man makes fiction: he invents stories, pretty enough, fantastical enough. But out of what does he make them up? Out of a few things in this great world which he has seen, and heard, and felt, just as he makes up his dreams. But who makes truth? Who makes facts? Who, but God?

Then truth is as much larger than fiction, as God is greater than man; as much larger as the whole universe is larger than the little corner of it that any man, even the greatest poet or philosopher, can see; and as much grander, and as much more beautiful, and as much more strange. For one is the whole, and the other is one, a few tiny scraps of the whole. The one is the work of God; the other is the work of man. Be sure that no man can ever fancy anything strange, unexpected, and curious, without finding if he had eyes to see, a hundred things around his feet more strange, more unexpected, more curious, actually ready-made already by God. You are fond of fairy tales, because they are fanciful, and like your dreams. My dear child, as your eyes open to the true fairy tale which Madam How can tell you all day long, nursery stories will seem to you poor and dull. All those feelings in you which your nursery tales call out,—imagination, wonder, awe, pity, and I trust too, hope and love—will be called out, I believe, by the Tale of all Tales, the true "Märchen allen Märchen," so much more fully and strongly and purely, that you will feel that novels and story-books are scarcely worth your reading, as long as you can read the great green book, of which every bud is a letter, and every tree a page.

Wonder if you will. You cannot wonder too much. That you might wonder all your life long, God put you into this wondrous world, and gave you that faculty of wonder which he has not given to the brutes; which is at once the mother of sound science, and a pledge of immortality in a world more wondrous even than this. But wonder at the right thing, not at the wrong; at the real miracles and prodigies, not at the sham. Wonder not at the world of man. Waste not your admiration, interest, hope on it, its pretty toys, gay fashions, fine clothes, tawdry luxuries, silly amusements. Wonder at the works of God. You will not, perhaps, take my advice yet. The world of man looks so pretty, that you will needs have your peep at it, and stare into its shop windows; and if you can, go to a few of its stage plays, and dance at a few of its balls. Ah—well—After a wild dream comes an uneasy wakening; and after too many sweet things, comes a sick headache. And one morning you will awake, I trust and pray, from the world of man to the world of God, and wonder where wonder is due, and worship where worship is due. You will awake like a child who has been at a pantomime over night, staring at the "fairy halls," which are all paint and canvas; and the "dazzling splendours," which are gas and oil; and the "magic transformations," which are done with ropes and pulleys; and the "brilliant elves," who are poor little children out of the next foul alley; and the harlequin and clown, who through all their fun are thinking wearily over the old debts which they must pay, and the hungry mouths at home which they must feed: and so, having thought it all wondrously glorious, and quite a fairy land, slips tired and stupid into bed, and wakes next morning to see the pure light shining in through the delicate frost-lace on the window-pane, and looks out over fields of virgin snow, and watches the rosy dawn and cloudless blue, and the great sun rising to the music of cawing rooks and piping stares, and says, "This is the true wonder. This is the true glory. The theatre last night was the fairy land of man; but this is the fairy land of God."

Chapter VII. The Chalk-Carts

What do you want to know about next? More about the caves in which the old savages lived,—how they were made, and how the curious things inside them got there, and so forth.

Well, we will talk about that in good time: but now—What is that coming down the hill?

Oh, only some chalk-carts.

CHALK-PIT

Only some chalk-carts? It seems to me that these chalk-carts are the very things we want; that if we follow them far enough—I do not mean with our feet along the public road, but with our thoughts along a road which, I am sorry to say, the public do not yet know much about—we shall come to a cave, and understand how a cave is made. Meanwhile, do not be in a hurry to say, "Only a chalk-cart," or only a mouse, or only a dead leaf. Chalk-carts, like mice, and dead leaves, and most other matters in the universe are very curious and odd things in the eyes of wise and reasonable people. Whenever I hear young men saying "only" this and "only" that, I begin to suspect them of belonging, not to the noble army of sages—much less to the most noble army of martyrs,—but to the ignoble army of noodles, who think nothing interesting or important but dinners, and balls, and races, and back-biting their neighbours; and I should be sorry to see you enlisting in that regiment when you grow up. But think—are not chalk-carts very odd and curious things? I think they are. To my mind, it is a curious question how men ever thought of inventing wheels; and, again, when they first thought of it. It is a curious question, too, how men ever found out that they could make horses work for them, and so began to tame them, instead of eating them, and a curious question (which I think we shall never get answered) when the first horse-tamer lived, and in what country. And a very curious, and, to me, a beautiful sight it is, to see those two noble horses obeying that little boy, whom they could kill with a single kick.

But, beside all this, there is a question, which ought to be a curious one to you (for I suspect you cannot answer it)—Why does the farmer take the trouble to send his cart and horses eight miles and more, to draw in chalk from Odiham chalk-pit?

Oh, he is going to put it on the land, of course. They are chalking the bit at the top of the next field, where the copse was grubbed.

But what good will he do by putting chalk on it? Chalk is not rich and fertile, like manure, it is altogether poor, barren stuff: you know that, or ought to know it. Recollect the chalk cuttings and banks on the railway between Basingstoke and Winchester—how utterly barren they are. Though they have been open these thirty years, not a blade of grass, hardly a bit of moss, has grown on them, or will grow, perhaps, for centuries.

Come, let us find out something about the chalk before we talk about the caves. The chalk is here, and the caves are not; and "Learn from the thing that lies nearest you" is as good a rule as "Do the duty which lies nearest you." Let us come into the grubbed bit, and ask the farmer—there he is in his gig.

Well, old friend, and how are you? Here is a little boy who wants to know why you are putting chalk on your field.

Does he then? If he ever tries to farm round here, he will have to learn for his first rule—No chalk, no wheat.

But why?

Why, is more than I can tell, young squire. But if you want to see how it comes about, look here at this freshly-grubbed land—how sour it is. You can see that by the colour of it—some black, some red, some green, some yellow, all full of sour iron, which will let nothing grow. After the chalk has been on it a year or two, those colours will have all gone out of it; and it will turn to a nice wholesome brown, like the rest of the field; and then you will know that the land is sweet, and fit for any crop. Now do you mind what I tell you, and then I'll tell you something more. We put on the chalk because, beside sweetening the land, it will hold water. You see, the land about here, though it is often very wet from springs, is sandy and hungry; and when we drain the bottom water out of it, the top water (that is, the rain) is apt to run through it too fast: and then it dries and burns up; and we get no plant of wheat, nor of turnips either. So we put on chalk to hold water, and keep the ground moist.

But how can these lumps of chalk hold water? They are not made like cups.

No: but they are made like sponges, which serves our turn better still. Just take up that lump, young squire, and you'll see water enough in it, or rather looking out of it, and staring you in the face.

Why! one side of the lump is all over thick ice.

So it is. All that water was inside the chalk last night, till it froze. And then it came squeezing out of the holes in the chalk in strings, as you may see it if you break the ice across. Now you may judge for yourself how much water a load of chalk will hold, even on a dry summer's day. And now, if you'll excuse me, sir, I must be off to market.

Was it all true that the farmer said?

Quite true, I believe. He is not a scientific man—that is, he does not know the chemical causes of all these things; but his knowledge is sound and useful, because it comes from long experience. He and his forefathers, perhaps for a thousand years and more, have been farming this country, reading Madam How's books with very keen eyes, experimenting and watching, very carefully and rationally; making mistakes often, and failing and losing their crops and their money; but learning from their mistakes, till their empiric knowledge, as it is called, helps them to grow sometimes quite as good crops as if they had learned agricultural chemistry.

What he meant by the chalk sweetening the land you would not understand yet, and I can hardly tell you; for chemists are not yet agreed how it happens. But he was right; and right, too, what he told you about the water inside the chalk, which is more important to us just now; for, if we follow it out, we shall surely come to a cave at last.

So now for the water in the chalk. You can see now why the chalk-downs at Winchester are always green, even in the hottest summer: because Madam How has put under them her great chalk sponge. The winter rains soak into it; and the summer heat draws that rain out of it again as invisible steam, coming up from below, to keep the roots of the turf cool and moist under the blazing sun.

You love that short turf well. You love to run and race over the Downs with your butterfly-net and hunt "chalk-hill blues," and "marbled whites," and "spotted burnets," till you are hot and tired; and then to sit down and look at the quiet little old city below, with the long cathedral roof, and the tower of St. Cross, and the gray old walls and buildings shrouded by noble trees, all embosomed among the soft rounded lines of the chalk-hills; and then you begin to feel very thirsty, and cry, "Oh, if there were but springs and brooks in the Downs, as there are at home!" But all the hollows are as dry as the hill tops. There is not a brook, or the mark of a watercourse, in one of them. You are like the Ancient Mariner in the poem, with

> "Water, water, every where,
> Nor any drop to drink."

To get that you must go down and down, hundreds of feet, to the green meadows through which silver Itchen glides toward the sea. There you stand upon the bridge, and watch the trout in water so crystal-clear that you see every weed and pebble as if you looked through air. If ever there was pure water, you think, that is pure. Is it so? Drink some. Wash your hands in it and try—You feel that the water is rough, hard (as they call it), quite different from the water at home, which feels as soft as velvet. What makes it so hard?

Because it is full of invisible chalk. In every gallon of that water there are, perhaps, fifteen grains of solid chalk, which was once inside the heart of the hills above. Day and night, year after year, the chalk goes down to the sea; and if there were such creatures as water-fairies—if it were true, as the old Greeks and Romans thought, that rivers were living things, with a Nymph who dwelt in each of them, and was its goddess or its queen—then, if your ears were opened to hear her, the Nymph of Itchen might say to you—

So child, you think that I do nothing but, as your sister says when she sings Mr. Tennyson's beautiful song,

> "I chatter over stony ways,
>   In little sharps and trebles,
> I bubble into eddying bays,
>   I babble on the pebbles."

Yes. I do that: and I love, as the Nymphs loved of old, men who have eyes to see my beauty, and ears to discern my song, and to fit their own song to it, and tell how

> "'I wind about, and in and out,
>   With here a blossom sailing,
> And here and there a lusty trout,
>   And here and there a grayling,
>
> "'And here and there a foamy flake
>   Upon me, as I travel
> With many a silvery waterbreak
>   Above the golden gravel,

> "'And draw them all along, and flow
> To join the brimming river,
> For men may come and men may go,
> But I go on for ever.'"

Yes. That is all true: but if that were all, I should not be let to flow on for ever, in a world where Lady Why rules, and Madam How obeys. I only exist (like everything else, from the sun in heaven to the gnat which dances in his beam) on condition of working, whether we wish it or not, whether we know it or not. I am not an idle stream, only fit to chatter to those who bathe or fish in my waters, or even to give poets beautiful fancies about me. You little guess the work I do. For I am one of the daughters of Madam How, and, like her, work night and day, we know not why, though Lady Why must know. So day by day, and night by night, while you are sleeping (for I never sleep), I carry, delicate and soft as I am, a burden which giants could not bear: and yet I am never tired. Every drop of rain which the south-west wind brings from the West Indian seas gives me fresh life and strength to bear my burden; and it has need to do so; for every drop of rain lays a fresh burden on me. Every root and weed which grows in every field; every dead leaf which falls in the highwoods of many a parish, from the Grange and Woodmancote round to Farleigh and Preston, and so to Brighton and the Alresford downs;—ay, every atom of manure which the farmers put on the land—foul enough then, but pure enough before it touches me— each of these, giving off a tiny atom of what men call carbonic acid, melts a tiny grain of chalk, and helps to send it down through the solid hill by one of the million pores and veins which at once feed and burden my springs. Ages on ages I have worked on thus, carrying the chalk into the sea. And ages on ages, it may be, I shall work on yet; till I have done my work at last, and leveled the high downs into a flat sea-shore, with beds of flint gravel rattling in the shallow waves.

She might tell you that; and when she had told you, you would surely think of the clumsy chalk-cart rumbling down the hill, and then of the graceful stream, bearing silently its invisible load of chalk; and see how much more delicate and beautiful, as well as vast and wonderful, Madam How's work is than that of man.

But if you asked the nymph why she worked on for ever, she could not tell you. For like the Nymphs of old, and the Hamadryads who lived, in trees, and Undine, and the little Sea-maiden, she would have no soul; no reason; no power to say why.

It is for you, who are a reasonable being, to guess why: or at least listen to me if I guess for you, and say, perhaps—I can only say perhaps—that chalk may be going to make layers of rich marl in the sea between England and France; and those marl-beds may be upheaved and grow into dry land, and be ploughed, and sowed, and reaped by a wiser race of men, in a better-ordered world than this: or the chalk may have even a nobler destiny before it. That may happen to it, which has happened already to many a grain of lime. It may be carried thousands of miles away to help in building up a coral reef (what that is I must tell you afterwards). That coral reef may harden into limestone beds. Those beds may be covered up, pressed, and, it may be, heated, till they crystallize into white marble: and out of it fairer statues be carved, and grander temples built, than the world has ever yet seen.

And if that is not the reason why the chalk is being sent into the sea, then there is another reason, and probably a far better one. For, as I told you at first, Lady Why's intentions are far wiser and better than our fancies; and she—like Him whom she obeys—is able to do exceeding abundantly, beyond all that we can ask or think.

But you will say now that we have followed the chalk-cart a long way, without coming to the cave.

You are wrong. We have come to the very mouth of the cave. All we have to do is to say—not "Open Sesame," like Ali Baba in the tale of the Forty Thieves—but some word or two which Madam Why will teach us, and forthwith a hill will open, and we shall walk in, and behold rivers and cascades underground, stalactite pillars and stalagmite statues, and all the wonders of the grottoes of Adelsberg, Antiparos, or Kentucky.

Am I joking? Yes, and yet no; for you know that when I joke I am usually most in earnest. At least, I am now.

But there are no caves in chalk?

No, not that I ever heard of. There are, though, in limestone, which is only a harder kind of chalk. Madam How could turn this chalk into hard limestone, I believe, even now; and in more ways than one: but in ways which would not be very comfortable or profitable for us Southern folk who live on it. I am afraid that—what between squeezing and heating—she would flatten us all out into phosphatic fossils, about an inch thick; and turn Winchester city into a "breccia" which would puzzle geologists a hundred thousand years hence. So we will hope that she will leave our chalk downs for the Itchen to wash gently away, while we talk about caves, and how Madam How scoops them out by water underground, just in the same way, only more roughly, as she melts the chalk.

Suppose, then, that these hills, instead of being soft, spongy chalk, were all hard limestone marble, like that of which the font in the church is made. Then the rainwater, instead of sinking through the chalk as now, would run over the ground down-hill, and if it came to a crack (a fault, as it is called) it would run down between the rock; and as it ran it would eat that hole wider and wider year by year, and make a swallow-hole—such as you may see in plenty if you ever go up Whernside, or any of the high hills in Yorkshire—unfathomable pits in the green turf, in which you may hear the water tinkling and trickling far, far underground.

And now, before we go a step farther, you may understand, why the bones of animals are so often found in limestone caves. Down such swallow-holes how many beasts must fall: either in hurry and fright, when hunted by lions and bears and such cruel beasts; or more often still in time of snow, when the holes are covered with drift; or, again, if they died on the open hill-sides, their bones might be washed in, in floods, along with mud and stones, and buried with them in the cave below; and beside that, lions and bears and hyenas might live in the caves below, as we know they did in some caves, and drag in bones through the caves' mouths; or, again, savages might live in that cave, and bring in animals to eat, like the wild beasts; and so those bones might be mixed up, as we know they were, with things which the savages had left behind—like flint tools or beads; and then the whole would be hardened, by the dripping of the limestone water, into a paste of breccia just like this in my drawer. But the bones of the savages themselves you would seldom or never find mixed in it—unless some one had fallen in by accident from above. And why? (For there is a Why? to that question: and not merely a How?) Simply because they were men; and because God has put into the hearts of all men, even of the lowest savages, some sort of reverence for those who are gone; and has taught them to bury, or in some other way take care of, their bones.

But how is the swallow-hole sure to end in a cave?

Because it cannot help making a cave for itself if it has time.

Think: and you will see that it must be so. For that water must run somewhere; and so it eats its way out between the beds of the rock, making underground galleries, and at last caves and lofty halls. For it always eats, remember, at the bottom of its channel, leaving the roof alone. So it eats, and eats, more in some places and less in others, according as the stone is harder or softer, and according to the different direction of the rock-beds (what we call their dip and strike); till at last it makes one of those wonderful caverns about which you are so fond of reading—such a cave as there actually is in the rocks of the mountain of Whernside, fed by the swallow-holes around the mountain-top; a cave hundreds of yards long, with halls, and lakes, and waterfalls, and curtains and festoons of stalactite which have dripped from the roof, and pillars of stalagmite which have been built up on the floor below. These stalactites (those tell me who have seen them) are among the most beautiful of all Madam How's work; sometimes like branches of roses or of grapes; sometimes like statues; sometimes like delicate curtains, and I know not what other beautiful shapes. I have never seen them, I am sorry to say, and therefore I cannot describe them. But they are all made in the same way; just in the same way as those little straight stalactites which you may have seen hanging, like icicles, in vaulted cellars, or under the arches of a bridge. The water melts more lime than it can carry, and drops some of it again, making fresh limestone grain by grain as it drips from the roof above; and fresh limestone again where it splashes on the floor below: till if it dripped long enough, the stalactite hanging from above would meet the stalagmite rising from below, and join in one straight round white graceful shaft, which would seem (but only seem) to support the roof of the cave. And out of that cave—though not always out of the mouth of it—will run a stream of water, which seems to you clear as crystal, though it is actually, like the Itchen at Winchester, full of lime; so full of lime, that it makes beds of fresh limestone, which are called travertine—which you may see in Italy, and Greece, and Asia Minor: or perhaps it petrifies, as you call it, the weeds in its bed, like that dropping-well at Knaresborough, of which you have often seen a picture. And the cause is this: the water is so full of lime, that it is forced to throw away some of it upon everything it touches, and so incrusts with stone—though it does not turn to stone—almost anything you put in it. You have seen, or ought to have seen, petrified moss and birds' nests and such things from Knaresborough Well: and now you know a little, though only a very little, of how the pretty toys are made.

Now if you can imagine for yourself (though I suppose a little boy cannot) the amount of lime which one of these subterranean rivers would carry away, gnawing underground centuries after centuries, day and night, summer and winter, then you will not be surprised at the enormous size of caverns which may be seen in different parts of the world—but always, I believe, in limestone rock. You would not be surprised (though you would admire them) at the caverns of Adelsberg, in Carniola (in the south of Austria, near the top of the Adriatic), which runs, I believe, for miles in length; and in the lakes of which, in darkness from its birth until its death, lives that strange beast, the Proteus a sort of long newt which never comes to perfection—I suppose for want of the genial sunlight which makes all things grow. But he is blind; and more, he keeps all his life the same feathery gills which newts have when they are babies, and which we have so often looked at through the microscope, to see the blood-globules run round and round inside. You would not wonder, either, at the Czirknitz Lake, near the same place, which at certain times of the year vanishes suddenly through chasms under water, sucking the fish down with it; and after a certain time boils suddenly up again from the depths, bringing back with it the fish, who have been swimming comfortably all the time in a subterranean lake; and bringing back, too (and, extraordinary as this story is, there is good reason to believe it true), live wild ducks who went down small and unfledged, and come back full-grown and fat, with water-weeds and small fish in their stomachs, showing they have had plenty to feed on underground. But—and this is the strangest part of the story, if true—they come up unfledged just as they went down, and are moreover blind from having been so long in darkness. After a while, however, folks say their eyes get right, their feathers grow, and they fly away like other birds.

Neither would you be surprised (if you recollect that Madam How is a very old lady indeed, and that some of her work is very old likewise) at that Mammoth Cave in Kentucky, the largest cave in the known world, through which you may walk nearly ten miles on end, and in which a hundred miles of gallery have been explored already, and yet no end found to the cave. In it (the guides will tell you) there are "226 avenues, 47 domes, 8 cataracts, 23 pits, and several rivers;" and if that fact is not very interesting to you (as it certainly is not to me) I will tell you something which ought to interest you: that this cave is so immensely old that various kinds of little animals, who have settled themselves in the outer parts of it, have had time to change their shape, and to become quite blind; so that blind fathers and mothers have blind children, generation after generation.

There are blind rats there, with large shining eyes which cannot see—blind landcrabs, who have the foot-stalks of their eyes (you may see them in any crab) still left; but the eyes which should be on the top of them are gone. There are blind fish, too, in the cave, and blind insects; for, if they have no use for their eyes in the dark, why should Madam How take the trouble to finish them off?

One more cave I must tell you of, to show you how old some caves must be, and then I must stop; and that is the cave of Caripé, in Venezuela, which is the most northerly part of South America. There, in the face of a limestone cliff, crested with enormous flowering trees, and festooned with those lovely creepers of which you have seen a few small ones in hothouses, there opens an arch as big as the west front of Winchester Cathedral, and runs straight in like a cathedral nave for more than 1400 feet. Out of it runs a stream; and along the banks of that stream, as far as the sunlight strikes in, grow wild bananas, and palms, and lords and ladies (as you call them), which are not, like ours, one foot, but many feet high. Beyond that the cave goes on, with subterranean streams, cascades, and halls, no man yet knows how far. A friend of mine last year went in farther, I believe, than any one yet has gone; but, instead of taking Indian torches made of bark and resin, or even torches made of Spanish wax, such as a brave bishop of those parts used once when he went in farther than any one before him, he took with him some of that beautiful magnesium light which you have seen often here at home. And in one place, when he lighted up the magnesium, he found himself in a hall full 300 feet high—higher far, that is, than the dome of St. Paul's—and a very solemn thought it was to him, he said, that he had seen what no other human being ever had seen; and that no ray of light had ever struck on that stupendous roof in all the ages since the making of the world. But if he found out something which he did not expect, he was disappointed in something which he did expect. For the Indians warned him of a hole in the floor which (they told him) was an unfathomable abyss. And lo and behold, when he turned the magnesium light upon it, the said abyss was just about eight feet deep. But it is no wonder that the poor Indians with their little smoky torches should make such mistakes; no wonder, too, that they should be afraid to enter far into those gloomy vaults; that they should believe that the souls of their ancestors live in that dark cave; and that they should say that when they die they will go to the Guacharos, as they call the birds that fly with doleful screams out of the cave to feed at night, and in again at daylight, to roost and sleep.

Now, it is these very Guacharo birds which are to me the most wonderful part of the story. The Indians kill and eat them for their fat, although they believe they have to do with evil spirits. But scientific men who have studied these birds will tell you that they are more wonderful than if all the Indians' fancies about them were true. They are great birds, more than three feet across the wings, somewhat like owls, somewhat like cuckoos, somewhat like goatsuckers; but, on the whole, unlike anything in the world but themselves; and instead of feeding on moths or mice, they feed upon hard dry fruits, which they pick off the trees after the set of sun. And wise men will tell you, that in making such a bird as that, and giving it that peculiar way of life, and settling it in that cavern, and a few more caverns in that part of the world, and therefore in making the caverns ready for them to live in, Madam How must have taken ages and ages, more than you can imagine or count.

But that is among the harder lessons which come in the latter part of Madam How's book. Children need not learn them yet; and they can never learn them, unless they master her alphabet, and her short and easy lessons for beginners, some of which I am trying to teach you now.

But I have just recollected that we are a couple of very stupid fellows. We have been talking all this time about chalk and limestone, and have forgotten to settle what they are, and how they were made. We must think of that next time. It will not do for us (at least if we mean to be scientific men) to use terms without defining them; in plain English, to talk about—we don't know what.

Chapter VIII. Madam How's Two Grandsons

You want to know, then, what chalk is? I suppose you mean what chalk is made of?

Yes. That is it.

That we can only help by calling in the help of a very great giant whose name is Analysis.

A giant?

Yes. And before we call for him I will tell you a very curious story about him and his younger brother, which is every word of it true.

Once upon a time, certainly as long ago as the first man, or perhaps the first rational being of any kind, was created, Madam How had two grandsons. The elder is called Analysis, and the younger Synthesis. As for who their father and mother were, there have been so many disputes on that question that I think children may leave it alone for the present. For my part, I believe that they are both, like St. Patrick, "gentlemen, and come of decent people;" and I have a great respect and affection for them both, as long as each keeps in his own place and minds his own business.

Now you must understand that, as soon as these two baby giants were born, Lady Why, who sets everything to do that work for which it is exactly fitted, set both of them their work. Analysis was to take to pieces everything he found, and find out how it was made. Synthesis was to put the pieces together again, and make something fresh out of them. In a word, Analysis was to teach men Science; and Synthesis to teach them Art.

But because Analysis was the elder, Madam How commanded Synthesis never to put the pieces together till Analysis had taken them completely apart. And, my child, if Synthesis had obeyed that rule of his good old grandmother's, the world would have been far happier, wealthier, wiser, and better than it is now.

But Synthesis would not. He grew up a very noble boy. He could carve, he could paint, he could build, he could make music, and write poems: but he was full of conceit and haste. Whenever his elder brother tried to do a little patient work in taking things to pieces, Synthesis snatched the work out of his hands before it was a quarter done, and began putting it together again to suit his own fancy, and, of course, put it together wrong. Then he went on to bully his elder brother, and locked him up in prison, and starved him, till for many hundred years poor Analysis never grew at all, but remained dwarfed, and stupid, and all but blind for want of light; while Synthesis, and all the hasty conceited people who followed him, grew stout and strong and tyrannous, and overspread the whole world, and ruled it at their will. But the fault of all the work of Synthesis was just this: that it would not work. His watches would not keep time, his soldiers would not fight, his ships would not sail, his houses would not keep the rain out. So every time he failed in his work he had to go to poor Analysis in his dungeon, and bully him into taking a thing or two to pieces, and giving him a few sound facts out of them, just to go on with till he came to grief again, boasting in the meantime that he and not Analysis had found out the facts. And at last he grew so conceited that he fancied he knew all that Madam How could teach him, or Lady Why either, and that he understood all things in heaven and earth; while it was not the real heaven and earth that he was thinking of, but a sham heaven and a sham earth, which he had built up out of his guesses and his own fancies.

And the more Synthesis waxed in pride, and the more he trampled upon his poor brother, the more reckless he grew, and the more willing to deceive himself. If his real flowers would not grow, he cut out paper flowers, and painted them and said that they would do just as well as natural ones. If his dolls would not work, he put strings and wires behind them to make them nod their heads and open their eyes, and then persuaded other people, and perhaps half-persuaded himself, that they were alive. If the hand of his weather-glass went down, he nailed it up to insure a fine day, and tortured, burnt, or murdered every one who said it did not keep up of itself. And many other foolish and wicked things he did, which little boys need not hear of yet.

But at last his punishment came, according to the laws of his grandmother, Madam How, which are like the laws of the Medes and Persians, and alter not, as you and all mankind will sooner or later find; for he grew so rich and powerful that he grew careless and lazy, and thought about nothing but eating and drinking, till people began to despise him more and more. And one day he left the dungeon of Analysis so ill guarded, that Analysis got out and ran away. Great was the hue and cry after him; and terribly would he have been punished had he been caught. But, lo and behold, folks had grown so disgusted with Synthesis that they began to take the part of Analysis. Poor men hid him in their cottages, and scholars in their studies. And when war arose about him,—and terrible wars did arise,—good kings, wise statesmen, gallant soldiers, spent their treasure and their lives in fighting for him. All honest folk welcomed him, because he was honest; and all wise folk used him, for, instead of being a conceited tyrant like Synthesis, he showed himself the most faithful, diligent, humble of servants, ready to do every man's work, and answer every man's questions. And among them all he got so well fed that he grew very shortly into the giant that he ought to have been all along; and was, and will be for many a year to come, perfectly able to take care of himself.

As for poor Synthesis, he really has fallen so low in these days, that one cannot but pity him. He now goes about humbly after his brother, feeding on any scraps that are thrown to him, and is snubbed and rapped over the knuckles, and told one minute to hold his tongue and mind his own business, and the next that he has no business at all to mind, till he has got into such a poor way that some folks fancy he will die, and are actually digging his grave already, and composing his epitaph. But they are trying to wear the bear's skin before the bear is killed; for Synthesis is not dead, nor anything like it; and he will rise up again some day, to make good friends with his brother Analysis, and by his help do nobler and more beautiful work than he has ever yet done in the world.

So now Analysis has got the upper hand; so much so that he is in danger of being spoilt by too much prosperity, as his brother was before him; in which case he too will have his fall; and a great deal of good it will do him. And that is the end of my story, and a true story it is.

Now you must remember, whenever you have to do with him, that Analysis, like fire, is a very good servant, but a very bad master. For, having got his freedom only of late years or so, he is, like young men when they come suddenly to be their own masters, apt to be conceited, and to fancy that he knows everything, when really he knows nothing, and can never know anything, but only knows about things, which is a very different matter. Indeed, nowadays he pretends that he can teach his old grandmother, Madam How, not only how to suck eggs, but to make eggs into the bargain; while the good old lady just laughs at him kindly, and lets him run on, because she knows he will grow wiser in time, and learn humility by his mistakes and failures, as I hope you will from yours.

However, Analysis is a very clever young giant, and can do wonderful work as long as he meddles only with dead things, like this bit of lime. He can take it to pieces, and tell you of what things it is made, or seems to be made; and take them to pieces again, and tell you what each of them is made of; and so on, till he gets conceited, and fancies that he can find out some one Thing of all things (which he calls matter), of which all other things are made; and some Way of all ways (which he calls force), by which all things are made: but when he boasts in that way, old Madam How smiles, and says, "My child, before you can say that, you must remember a hundred things which you are forgetting, and learn a hundred thousand things which you do not know;" and then she just puts her hand over his eyes, and Master Analysis begins groping in the dark, and talking the saddest nonsense. So beware of him, and keep him in his own place, and to his own work, or he will flatter you, and get the mastery of you, and persuade you that he can teach you a thousand things of which he knows no more than he does why a duck's egg never hatches into a chicken. And remember, if Master Analysis ever grows saucy and conceited with you, just ask him that last riddle, and you will shut him up at once.

And why?

Because Analysis can only explain to you a little about dead things, like stones—inorganic things as they are called. Living things—organisms, as they are called—he cannot explain to you at all. When he meddles with them, he always ends like the man who killed his goose to get the golden eggs. He has to kill his goose, or his flower, or his insect, before he can analyze it; and then it is not a goose, but only the corpse of a goose; not a flower, but only the dead stuff of the flower.

And therefore he will never do anything but fail, when he tries to find out the life in things. How can he, when he has to take the life out of them first? He could not even find out how a plum-pudding is made by merely analyzing it. He might part the sugar, and the flour, and the suet; he might even (for he is very clever, and very patient too, the more honour to him) take every atom of sugar out of the flour with which it had got mixed, and every atom of brown colour which had got out of the plums and currants into the body of the pudding, and then, for aught I know, put the colouring matter back again into the plums and currants; and then, for aught I know, turn the boiled pudding into a raw one again,—for he is a great conjurer, as Madam How's grandson is bound to be: but yet he would never find out how the pudding was made, unless some one told him the great secret which the sailors in the old story forgot—that the cook boiled it in a cloth.

This is Analysis's weak point—don't let it be yours—that in all his calculations he is apt to forget the cloth, and indeed the cook likewise. No doubt he can analyze the matter of things: but he will keep forgetting that he cannot analyze their form.

Do I mean their shape?

No, my child; no. I mean something which makes the shape of things, and the matter of them likewise, but which folks have lost sight of nowadays, and do not seem likely to get sight of again for a few hundred years. So I suppose that you need not trouble your head about it, but may just follow the fashions as long as they last.

About this piece of lime, however, Analysis can tell us a great deal. And we may trust what he says, and believe that he understands what he says.

Why?

Think now. If you took your watch to pieces, you would probably spoil it for ever; you would have perhaps broken, and certainly mislaid, some of the bits; and not even a watchmaker could put it together again. You would have analyzed the watch wrongly. But if a watchmaker took it to pieces then any other watchmaker could put it together again to go as well as ever, because they both understand the works, how they fit into each other, and what the use and the power of each is. Its being put together again rightly would be a proof that it had been taken to pieces rightly.

And so with Master Analysis. If he can take a thing to pieces so that his brother Synthesis can put it together again, you may be sure that he has done his work rightly.

Now he can take a bit of chalk to pieces, so that it shall become several different things, none of which is chalk, or like chalk at all. And then his brother Synthesis can put them together again, so that they shall become chalk, as they were before. He can do that very nearly, but not quite. There is, in every average piece of chalk, something which he cannot make into chalk again when he has once unmade it.

What that is I will show you presently; and a wonderful tale hangs thereby. But first we will let Analysis tell us what chalk is made of, as far as he knows.

He will say—Chalk is carbonate of lime.

But what is carbonate of lime made of?

Lime and carbonic acid.

And what is lime?

The oxide of a certain metal, called calcium.

What do you mean?

That quicklime is a certain metal mixed with oxygen gas; and slacked lime is the same, mixed with water.

So lime is a metal. What is a metal? Nobody knows.

And what is oxygen gas? Nobody knows.

Well, Analysis, stops short very soon. He does not seem to know much about the matter.

Nay, nay, you are wrong there. It is just "about the matter" that he does know, and knows a great deal, and very accurately; what he does not know is the matter itself. He will tell you wonderful things about oxygen gas—how the air is full of it, the water full of it, every living thing full of it; how it changes hard bright steel into soft, foul rust; how a candle cannot burn without it, or you live without it. But what it is he knows not.

Will he ever know?

That is Lady Why's concern, and not ours. Meanwhile he has a right to find out if he can. But what do you want to ask him next?

What? Oh! What carbonic acid is. He can tell you that. Carbon and oxygen gas.

But what is carbon?

Nobody knows.

Why, here is this stupid Analysis at fault again.

Nay, nay, again. Be patient with him. If he cannot tell you what carbon is, he can tell you what is carbon, which is well worth knowing. He will tell you, for instance, that every time you breathe or speak, what comes out of your mouth is carbonic acid; and that, if your breath comes on a bit of slacked lime, it will begin to turn it back into the chalk from which it was made; and that, if your breath comes on the leaves of a growing plant, that leaf will take the carbon out of it, and turn it into wood. And surely that is worth knowing,—that you may be helping to make chalk, or to make wood, every time you breathe.

Well; that is very curious.

But now, ask him, What is carbon? And he will tell you, that many things are carbon. A diamond is carbon; and so is blacklead; and so is charcoal and coke, and coal in part, and wood in part.

What? Does Analysis say that a diamond and charcoal are the same thing?

Yes.

Then his way of taking things to pieces must be a very clumsy one, if he can find out no difference between diamond and charcoal.

Well, perhaps it is: but you must remember that, though he is very old—as old as the first man who ever lived—he has only been at school for the last three hundred years or so. And remember, too, that he is not like you, who have some one else to teach you. He has had to teach himself, and find out for himself, and make his own tools, and work in the dark besides. And I think it is very much to his credit that he ever found out that diamond and charcoal were the same things. You would never have found it out for yourself, you will agree.

No: but how did he do it?

He taught a very famous chemist, Lavoisier, about ninety years ago, how to burn a diamond in oxygen—and a very difficult trick that is; and Lavoisier found that the diamond when burnt turned almost entirely into carbonic acid and water, as blacklead and charcoal do; and more, that each of them turned into the same quantity of carbonic acid, And so he knew, as surely as man can know anything, that all these things, however different to our eyes and fingers, are really made of the same thing,—pure carbon.

But what makes them look and feel so different?

That Analysis does not know yet. Perhaps he will find out some day; for he is very patient, and very diligent, as you ought to be. Meanwhile, be content with him: remember that though he cannot see through a milestone yet, he can see farther into one than his neighbours. Indeed his neighbours cannot see into a milestone at all, but only see the outside of it, and know things only by rote, like parrots, without understanding what they mean and how they are made.

So now remember that chalk is carbonate of lime, and that it is made up of three things, calcium, oxygen, and carbon; and that therefore its mark is $CaCO_3$, in Analysis's language, which I hope you will be able to read some day.

But how is it that Analysis and Synthesis cannot take all this chalk to pieces, and put it together again?

Look here; what is that in the chalk?

Oh! a shepherd's crown, such as we often find in the gravel, only fresh and white.

Well; you know what that was once. I have often told you:—a live sea-egg, covered with prickles, which crawls at the bottom of the sea.

Well, I am sure that Master Synthesis could not put that together again: and equally sure that Master Analysis might spend ages in taking it to pieces, before he found out how it was made. And—we are lucky to-day, for this lower chalk to the south has very few fossils in it—here is something else which is not mere carbonate of lime. Look at it.

A little cockle, something like a wrinkled hazel-nut.

No; that is no cockle. Madam How invented that ages and ages before she thought of cockles, and the animal which lived inside that shell was as different from a cockle-animal as a sparrow is from a dog. That is a Terebratula, a gentleman of a very ancient and worn-out family. He and his kin swarmed in the old seas, even as far back as the time when the rocks of the Welsh mountains were soft mud; as you will know when you read that great book of Sir Roderick Murchison's, *Siluria*. But as the ages rolled on, they got fewer and fewer, these Terebratulæ; and now there are hardly any of them left; only six or seven sorts are left about these islands, which cling to stones in deep water; and the first time I dredged two of them out of Loch Fyne, I looked at them with awe, as on relics from another world, which had lasted on through unnumbered ages and changes, such as one's fancy could not grasp.

But you will agree that, if Master Analysis took that shell to pieces, Master Synthesis would not be likely to put it together again; much less to put it together in the right way, in which Madam How made it.

And what was that?

By making a living animal, which went on growing, that is, making itself; and making, as it grew, its shell to live in. Synthesis has not found out yet the first step towards doing that; and, as I believe, he never will.

But there would be no harm in his trying?

Of course not. Let everybody try to do everything they fancy. Even if they fail, they will have learnt at least that they cannot do it.

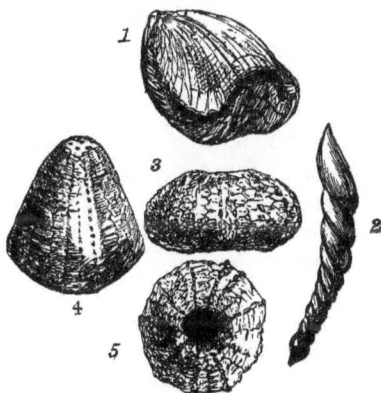

1. TEREBRATULA.   2. DENTALINA.   3, 5. CIDARIS DIADEMA.
4. GALERITES.

But now—and this is a secret which you would never find out for yourself, at least without the help of a microscope—the greater part of this lump of chalk is made up of things which neither Analysis can perfectly take to pieces, nor Synthesis put together again. It is made of dead organisms, that is, things which have been made by living creatures. If you washed and brushed that chalk into powder, you would find it full of little things like the Dentalina in this drawing, and many other curious forms. I will show you some under the microscope one day.

They are the shells of animals called Foraminifera, because the shells of some of them are full of holes, through which they put out tiny arms. So small they are and so many, that there may be, it is said, forty thousand of them in a bit of chalk an inch every way. In numbers past counting, some whole, some broken, some ground to the finest powder, they make up vast masses of England, which are now chalk downs; and in some foreign countries they make up whole mountains. Part of the building stone of the Great Pyramid in Egypt is composed, I am told, entirely of them.

And how did they get into the chalk?

Ah! How indeed? Let us think. The chalk must have been laid down at the bottom of a sea, because there are sea-shells in it. Besides, we find little atomies exactly like these alive now in many seas; and therefore it is fair to suppose these lived in the sea also.

Besides, they were not washed into the chalk by any sudden flood. The water in which they settled must have been quite still, or these little delicate creatures would have been ground into powder—or rather into paste. Therefore learned men soon made up their minds that these things were laid down at the bottom of a deep sea, so deep that neither wind, nor tide, nor currents could stir the everlasting calm.

Ah! it is worth thinking over, for it shows how shrewd a giant Analysis is, and how fast he works in these days, now that he has got free and well fed;—worth thinking over, I say, how our notions about these little atomies have changed during the last forty years.

We used to find them sometimes washed up among the sea-sand on the wild Atlantic coast; and we were taught, in the days when old Dr. Turton was writing his book on British shells at Bideford, to call them Nautili, because their shells were like Nautilus shells. Men did not know then that the animal which lives in them is no more like a Nautilus animal than it is like a cow.

For a Nautilus, you must know, is made like a cuttlefish, with eyes, and strong jaws for biting, and arms round them; and has a heart, and gills, and a stomach; and is altogether a very well-made beast, and, I suspect, a terrible tyrant to little fish and sea-slugs, just as the cuttlefish is. But the creatures which live in these little shells are about the least finished of Madam How's works. They have neither mouth nor stomach, eyes nor limbs. They are mere live bags full of jelly, which can take almost any shape they like, and thrust out arms—or what serve for arms—through the holes in their shells, and then contract them into themselves again, as this Globigerina does. What they feed on, how they grow, how they make their exquisitely-formed shells, whether, indeed, they are, strictly speaking, animals or vegetables, Analysis has not yet found out. But when you come to read about them, you will find that they, in their own way, are just as wonderful and mysterious as a butterfly or a rose; and just as necessary, likewise, to Madam How's work; for out of them, as I told you, she makes whole sheets of down, whole ranges of hills.

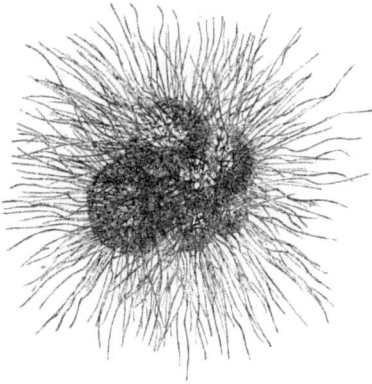

GLOBIGERINA

No one knew anything, I believe, about them, save that two or three kinds of them were found in chalk, till a famous Frenchman, called D'Orbigny, just thirty years ago, told the world how he had found many beautiful fresh kinds; and, more strange still, that some of these kinds were still alive at the bottom of the Adriatic, and of the harbour of Alexandria, in Egypt.

Then in 1841 a gentleman named Edward Forbes,—now with God—whose name will be for ever dear to all who love science, and honour genius and virtue,—found in the Ægean Sea "a bed of chalk," he said, "full of Foraminifera, and shells of Pteropods," forming at the bottom of the sea.

And what are Pteropods?

What you might call sea-moths (though they are not really moths), which swim about on the surface of the water, while the right-whales suck them in tens of thousands into the great whalebone net which fringes their jaws. Here are drawings of them. 1. Limacina (on which the whales feed); and 2. Hyalea, a lovely little thing in a glass shell, which lives in the Mediterranean.

But since then strange discoveries have been made, especially by the naval officers who surveyed the bottom of the great Atlantic Ocean before laying down the electric cable between Ireland and America. And this is what they found:

That at the bottom of the Atlantic were vast plains of soft mud, in some places 2500 fathoms (15,000 feet) deep; that is, as deep as the

Alps are high. And more: they found out, to their surprise, that the oozy mud of the Atlantic floor was made up almost entirely of just the same atomies as make up our chalk, especially globigerinas; that, in fact, a vast bed of chalk was now forming at the bottom of the Atlantic, with living shells and sea-animals of the most brilliant colours crawling about on it in black darkness, and beds of sponges growing out of it, just as the sponges grew at the bottom of the old chalk ocean, and were all, generation after generation, turned into flints.

And, for reasons which you will hardly understand, men are beginning now to believe that the chalk has never ceased to be made, somewhere or other, for many thousand years, ever since the Winchester Downs were at the bottom of the sea: and that "the Globigerina-mud is not merely *a* chalk formation, but a continuation of *the* chalk formation, so *that we may be said to be still living in the age of Chalk.*"[1] Ah, my little man, what would I not give to see you, before I die, add one such thought as that to the sum of human knowledge!

So there the little creatures have been lying, making chalk out of the lime in the sea-water, layer over layer, the young over the old, the dead over the living, year after year, age after age—for how long?

Who can tell? How deep the layer of new chalk at the bottom of the Atlantic is, we can never know. But the layer of live atomies on it is not an inch thick, probably not a tenth of an inch. And if it grew a tenth of an inch a year, or even a whole inch, how many years must it have

---

[1] I could not resist the temptation of quoting this splendid generalization from Dr. Carpenter's Preliminary Report of the Dredging Operations of H.M.S. "Lightening," 1868. He attributes it, generously, to his colleague, Dr. Wyville Thomson. Be it whose it may, it will mark (as will probably the whole Report when completed) a new era in Bio-Geology.

taken to make the chalk of our downs, which is in some parts 1300 feet thick? How many inches are there in 1300 feet? Do that sum, and judge for yourself.

One difference will be found between the chalk now forming at the bottom of the ocean, if it ever become dry land, and the chalk on which you tread on the downs. The new chalk will be full of the teeth and bones of whales—warm-blooded creatures, who suckle their young like cows, instead of laying eggs, like birds and fish. For there were no whales in the old chalk ocean; but our modern oceans are full of cachalots, porpoises, dolphins, swimming in shoals round any ship; and their bones and teeth, and still more their ear-bones, will drop to the bottom as they die, and be found, ages hence, in the mud which the live atomies make, along with wrecks of mighty ships

"Great anchors, heaps of pearl,"

and all that man has lost in the deep seas. And sadder fossils yet, my child, will be scattered on those white plains:—

"To them the love of woman hath gone down,
    Dark roll their waves o'er manhood's noble head.
O'er youth's bright locks, and beauty's flowing crown;
    Yet shall they hear a voice, 'Restore the dead.'
Earth shall reclaim her precious things from thee.
    Give back the dead, thou Sea!"

Chapter IX. The Coral-Reef

Now you want to know what I meant when I talked of a bit of lime going out to sea, and forming part of a coral island, and then of a limestone rock, and then of a marble statue. Very good. Then look at this stone.

What a curious stone! Did it come from any place near here?

No. It came from near Dudley, in Staffordshire, where the soils are worlds on worlds older than they are here, though they were made in the same way as these and all other soils. But you are not listening to me.

Why, the stone is full of shells, and bits of coral; and what are these wonderful things coiled and tangled together, like the snakes in Medusa's hair in the picture? Are they snakes?

If they are, then they must be snakes who have all one head; for see, they are joined together at their larger ends; and snakes which are branched, too, which no snake ever was.

Yes. I suppose they are not snakes. And they grow out of a flower, too; and it has a stalk, jointed, too, as plants sometimes are; and as fishes' backbones are too. Is it a petrified plant or flower?

No; though I do not deny that it looks like one. The creature most akin to it which you ever saw is a star-fish.

What! one of the red star-fishes which one finds on the beach? Its arms are not branched.

No. But there are star-fishes with branched arms still in the sea. You know that pretty book (and learned book, too), Forbes's *British Star-fishes*? You like to look it through for the sake of the vignettes,— the mermaid and her child playing in the sea.

Oh yes, and the kind bogie who is piping while the sandstars dance; and the other who is trying to pull out the star-fish which the oyster has caught.

Yes. But do you recollect the drawing of the Medusa's head, with its curling arms, branched again and again without end? Here it is. No, you shall not look at the vignettes now. We must mind business. Now look at this one; the Feather-star, with arms almost like fern-fronds. And in foreign seas there are many other branched star-fish beside.

But they have no stalks?

Do not be too sure of that. This very feather-star, soon after it is born, grows a tiny stalk, by which it holds on to corallines and sea-weeds; and it is not till afterwards that it breaks loose from that stalk, and swims away freely into the wide water. And in foreign seas there are several star-fish still who grow on stalks all their lives, as this fossil one did.

How strange that a live animal should grow on a stalk, like a flower!

Not quite like a flower. A flower has roots, by which it feeds in the soil. These things grow more like sea-weeds, which have no roots, but only hold on to the rock by the foot of the stalk, as a ship holds on by her anchor. But as for its being strange that live animals should grow on stalks, if it be strange it is common enough, like many far stranger things. For under the water are millions on millions of creatures, spreading for miles on miles, building up at last great reefs of rocks, and whole islands, which all grow rooted first to the rock, like sea-weeds; and what is more, they grow, most of them, from one common root, branching again and again, and every branchlet bearing hundreds of living creatures, so that the whole creation is at once one creature and many creatures. Do you not understand me?

No.

Then fancy to yourself a bush like that hawthorn bush, with numberless blossoms, and every blossom on that bush a separate living thing, with its own mouth, and arms, and stomach, budding and growing fresh live branches and fresh live flowers, as fast as the old ones die: and then you will see better what I mean.

How wonderful!

Yes; but not more wonderful than your finger, for it, too, is made up of numberless living things.

My finger made of living things?

What else can it be? When you cut your finger, does not the place heal?

Of course.

And what is healing but growing again? And how could the atoms of your fingers grow, and make fresh skin, if they were not each of them alive? There, I will not puzzle you with too much at once; you will know more about all that some day. Only remember now, that there is nothing wonderful in the world outside you but has its counterpart of something just as wonderful, and perhaps more wonderful, inside you. Man is the microcosm, the little world, said the philosophers of old; and philosophers nowadays are beginning to see that their old guess is actual fact and true.

But what are these curious sea-creatures called, which are animals, yet grow like plants?

They have more names than I can tell you, or you remember. Those which helped to make this bit of stone are called coral-insects:

but they are not really insects, and are no more like insects than you
are. Coral-polypes is the best name for them, because they have arms
round their mouths, something like a cuttlefish, which the ancients
called Polypus. But the animal which you have seen likest to most of
them is a sea-anemone.

Look now at this piece of fresh coral—for coral it is, though not
like the coral which your sister wears in her necklace. You see it is full
of pipes; in each of those pipes has lived what we will call, for the time
being, a tiny sea-anemone, joined on to his brothers by some sort of
flesh and skin; and all of them together have built up, out of the lime in
the sea-water, this common house, or rather town, of lime.

But is it not strange and wonderful?

Of course it is: but so is everything when you begin to look into it;
and if I were to go on, and tell you what sort of young ones these coral-
polypes have, and what becomes of them, you would hear such
wonders, that you would be ready to suspect that I was inventing
nonsense, or talking in my dreams. But all that belongs to Madam
How's deepest book of all, which is called the BOOK OF KIND: the book
which children cannot understand, and in which only the very wisest
men are able to spell out a few words, not knowing, and of course not
daring to guess, what wonder may come next.

Now we will go back to our stone, and talk about how it was made,
and how the stalked star-fish, which you mistook for a flower, ever got
into the stone.

Then do you think me silly for fancying that a fossil star-fish was a
flower?

I should be silly if I did. There is no silliness in not knowing what you cannot know. You can only guess about new things, which you have never seen before, by comparing them with old things, which you have seen before; and you had seen flowers, and snakes, and fishes' backbones, and made a very fair guess from them. After all, some of these stalked star-fish are so like flowers, lilies especially, that they are called Encrinites; and the whole family is called Crinoids, or lily-like creatures, from the Greek work *krinon*, a lily; and as for corals and corallines, learned men, in spite of all their care and shrewdness, made mistake after mistake about them, which they had to correct again and again, till now, I trust, they have got at something very like the truth. No, I shall only call you silly if you do what some little boys are apt to do—call other boys, and, still worse, servants or poor people, silly for not knowing what they cannot know.

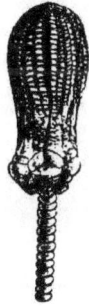

But are not poor people often very silly about animals and plants? The boys at the village school say that slowworms are poisonous; is not that silly?

Not at all. They know that adders bite, and so they think that slowworms bite too. They are wrong; and they must be told that they are wrong, and scolded if they kill a slowworm. But silly they are not.

But is it not silly to fancy that swallows sleep all the winter at the bottom of the pond?

I do not think so. The boys cannot know where the swallows go; and if you told them—what is true—that the swallows find their way every autumn through France, through Spain, over the Straits of Gibraltar, into Morocco, and some, I believe, over the great desert of Zahara into Negroland: and if you told them—what is true also—that the young swallows actually find their way into Africa without having been along the road before; because the old swallows go south a week or two first, and leave the young ones to guess out the way for themselves: if you told them that, then they would have a right to say, "Do you expect us to believe that? That is much more wonderful than that the swallows should sleep in the pond."

But is it?

Yes; to them. They know that bats and dormice and other things sleep all the winter; so why should not swallows sleep? They see the swallows about the water, and often dipping almost into it. They know that fishes live under water, and that many insects—like May-flies and caddis-flies and water-beetles—live sometimes in the water, sometimes in the open air; and they cannot know—you do not know—what it is which prevents a bird's living under water. So their guess is really a very fair one; no more silly than that of the savages, who when they first saw the white men's ships, with their huge sails, fancied they were enormous sea-birds; and when they heard the cannons fire, said that the ships spoke in thunder and lightning. Their guess was wrong, but not silly; for it was the best guess they could make.

But I do know of one old woman who was silly. She was a boy's nurse, and she gave the boy a thing which she said was one of the snakes which St. Hilda turned into stone; and told him that they found plenty of them at Whitby, where she was born, all coiled up; but what was very odd, their heads had always been broken of. And when he took it, to his father, he told him it was only a fossil shell—an Ammonite. And he went back and laughed at his nurse, and teased her till she was quite angry.

Then he was very lucky that she did not box his ears, for that was what he deserved. I dare say that, though his nurse had never heard of Ammonites, she was a wise old dame enough, and knew a hundred things which he did not know, and which were far more important than Ammonites, even to him.

How?

Because if she had not known how to nurse him well, he would perhaps have never grown up alive and strong. And if she had not known how to make him obey and speak the truth, he might have grown up a naughty boy.

But was she not silly?

No. She only believed what the Whitby folk, I understand, have some of them believed for many hundred years. And no one can be blamed for thinking as his forefathers did, unless he has cause to know better.

Surely she might have known better?

How? What reason could she have to believe the Ammonite was a shell? It is not the least like cockles, or whelks, or any shell she ever saw.

What reason either could she have to guess that Whitby cliff had once been coral-mud, at the bottom of the sea? No more reason, my dear child, than you would have to guess that this stone had been coral-mud likewise, if I did not teach you so,—or rather, try to make you teach yourself so.

No. I say it again. If you wish to learn, I will only teach you on condition that you do not laugh at, or despise, those good and honest and able people who do not know or care about these things, because they have other things to think of: like old John out there ploughing. He would not believe you—he would hardly believe me—if we told him that this stone had been once a swarm of living things, of exquisite shapes and glorious colours. And yet he can plough and sow, and reap and mow, and fell and strip, and hedge and ditch, and give his neighbours sound advice, and take the measure of a man's worth from ten minutes' talk, and say his prayers, and keep his temper, and pay his debts,—which last three things are more than a good many folks can do who fancy themselves a whole world wiser than John in the smock-frock.

Oh, but I want to hear about the exquisite shapes and glorious colours.

Of course you do, little man. A few fine epithets take your fancy far more than a little common sense and common humility; but in that you are no worse than some of your elders. So now for the exquisite shapes and glorious colours. I have never seen them; though I trust to see them ere I die. So what they are like I can only tell from what I have learnt from Mr. Darwin, and Mr. Wallace, and Mr. Jukes, and Mr. Gosse, and last, but not least, from one whose soul was as beautiful as his face, Lucas Barrett,—too soon lost to science,—who was drowned in exploring such a coral-reef as this stone was once.

Then there are such things alive now?

Yes, and no. The descendants of most of them live on, altered by time, which alters all things; and from the beauty of the children we can guess at the beauty of their ancestors; just as from the coral-reefs which exist now we can guess how the coral-reefs of old were made. And that this stone was once part of a coral-reef the corals in it prove at first sight.

And what is a coral-reef like?

You have seen the room in the British Museum full of corals, madrepores, brain-stones, corallines, and sea-ferns?

Oh yes.

Then fancy all those alive. Not as they are now, white stone: but covered in jelly; and out of every pore a little polype, like a flower, peeping out. Fancy them of every gaudy colour you choose. No bed of flowers, they say, can be more brilliant than the corals, as you look down on them through the clear sea. Fancy, again, growing among them and crawling over them, strange sea-anemones, shells, star-fish, sea-slugs, and sea-cucumbers with feathery gills, crabs, and shrimps, and hundreds of other animals, all as strange in shape, and as brilliant in colour. You may let your fancy run wild. Nothing so odd, nothing so gay, even entered your dreams, or a poet's, as you may find alive at the bottom of the sea, in the live flower-gardens of the sea-fairies.

There will be shoals of fish, too, playing in and out, as strange and gaudy as the rest,—parrot-fish who browse on the live coral with their beak-like teeth, as cattle browse on grass; and at the bottom, it may be, larger and uglier fish, who eat the crabs and shell-fish, shells and all, grinding them up as a dog grinds a bone, and so turning shells and corals into fine soft mud, such as this stone is partly made of.

But what happens to all the delicate little corals if a storm comes on?

What, indeed? Madam How has made them so well and wisely, that, like brave and good men, the more trouble they suffer the stronger they are. Day and night, week after week, the trade-wind blows upon them, hurling the waves against them in furious surf, knocking off great lumps of coral, grinding them to powder, throwing them over the reef into the shallow water inside. But the heavier the surf beats upon them, the stronger the polypes outside grow, repairing their broken houses, and building up fresh coral on the dead coral below, because it is in the fresh sea-water that beats upon the surf that they find most lime with which to build. And as they build they form a barrier against the surf, inside of which, in water still as glass, the weaker and more delicate things can grow in safety, just as these very Encrinites may have grown, rooted in the lime-mud, and waving their slender arms at the bottom of the clear lagoon. Such mighty builders are these little coral polypes, that all the works of men are small compared with theirs. One single reef, for instance, which is entirely made by them, stretches along the north-east coast of Australia for nearly a thousand miles. Of this you must read some day in Mr. Jukes's *Voyage of H.M.S. "Fly."* Every island throughout a great part of the Pacific is fringed round each with its coral-reef, and there are hundreds of islands of strange shapes, and of Atolls, as they are called, or ring-islands, which are composed entirely of coral, and of nothing else.

A ring-island? How can an island be made in the shape of a ring?

BIRGUS LATRO.

Ah! it was a long time before men found out that riddle. Mr. Darwin was the first to guess the answer, as he has guessed many an answer beside. These islands are each a ring, or nearly a ring of coral, with smooth shallow water inside: but their outsides run down, like a mountain wall, sheer into seas hundreds of fathoms deep. People used to believe, and reasonably enough, that the coral polypes began to build up the islands from the very bottom of the deep sea. But that would not account for the top of them being of the shape of a ring; and in time it was found out that the corals would not build except in shallow water, twenty or thirty fathoms deep at most, and men were at their wits' ends to find out the riddle. Then said Mr. Darwin, "Suppose one of those beautiful South Sea Islands, like Tahiti, the Queen of Isles, with its ring of coral-reef all round its shore, began sinking slowly under the sea. The land, as it sunk, would be gone for good and all: but the coral-reef round it would not, because the coral polypes would build up and up continually upon the skeletons of their dead parents, to get to the surface of the water, and would keep close to the top outside, however much the land sunk inside; and when the island had sunk completely beneath the sea, what would be left? What must be left but a ring of coral reef, around the spot where the last mountain peak of the island sank beneath the sea?" And so Mr. Darwin explained the shapes of hundreds of coral islands in the Pacific; and proved, too, some strange things besides (he proved, and other men, like Mr. Wallace, whose excellent book on the East Indian islands you must read some day, have proved in other ways) that there was once a great continent, joined perhaps to Australia and to New Guinea, in the Pacific Ocean, where is now nothing but deep sea, and coral-reefs which mark the mountain ranges of that sunken world.

But how does the coral ever rise above the surface of the water and turn into hard stone?

Of course the coral polypes cannot build above the high-tide mark; but the surf which beats upon them piles up their broken fragments just as a sea-beach is piled up, and hammers them together with that water hammer which is heavier and stronger than any you have ever seen in a smith's forge. And then, as is the fashion of lime, the whole mass sets and becomes hard, as you may see mortar set; and so you have a low island a few feet above the sea. Then sea-birds come to it, and rest and build; and seeds are floated thither from far lands; and among them almost always the cocoa-nut, which loves to grow by the sea-shore, and groves of cocoa palms grow up upon the lonely isle. Then, perhaps, trees and bushes are drifted thither before the trade-wind; and entangled in their roots are seeds of other plants, and eggs or cocoons of insects; and so a few flowers and a few butterflies and beetles set up for themselves upon the new land. And then a bird or two, caught in a storm and blown away to sea finds shelter in the cocoa-grove; and so a little new world is set up, in which (you must remember always) there are no four-footed beasts, nor snakes, nor lizards, nor frogs, nor any animals that cannot cross the sea. And on some of those islands they may live (indeed there is reason to believe they have lived), so long, that some of them have changed their forms, according to the laws of Madam How, who sooner or later fits each thing exactly for the place in which it is meant to live, till upon some of them you may find such strange and unique creatures as the famous cocoa-nut crab, which learned men call *Birgus latro*. A great crab he is, who walks upon the tips of his toes a foot high above the ground. And because he has often nothing to eat but cocoa-nuts, or at least they are the best things he can find, cocoa-nuts he has learned to eat, and after a fashion which it would puzzle you to imitate. Some say that he climbs up the stems of the cocoa-nut trees, and pulls the fruit down for himself; but that, it seems, he does not usually do. What he does is this: when he finds a fallen cocoa-nut, he begins tearing away the thick husk and fibre with his strong claws; and he knows perfectly well which end to tear it from, namely, from the end where the three eye-holes are, which you call the monkey's face, out of one of which you know, the young cocoa-nut tree would burst forth. And when he has got to the eye-holes, he hammers through one of them with the point of his heavy claw. So far, so good: but how is he to get the meat out? He cannot put his claw in. He has no proboscis like a butterfly to insert and suck with. He is as far off from his dinner as the fox was when the stork offered him a feast in a long-necked jar. What then do you think he does? He turns himself round, puts in a pair of his hind pincers, which are very slender, and with them scoops the meat out of the cocoa-nut, and so puts his dinner into his mouth with his hind feet. And even the cocoa-nut husk he does not waste; for he lives in deep burrows which he makes like a rabbit; and being a luxurious crab, and liking to sleep soft in spite of his hard shell, he lines them with a quantity of cocoa-nut fibre, picked out clean and fine, just as if he was going to make cocoa-nut matting of it. And being also a clean crab, as I hope you are a clean little boy, he goes down to the sea every night to have his bath and moisten his gills, and so lives happy all his days, and gets so fat in his old age that he carries about his body nearly a quart of pure oil.

That is the history of the cocoa-nut crab. And if any one tells me that that crab acts only on what is called "instinct"; and does not think and reason, just as you and I think and reason, though of course not in words as you and I do: then I shall be inclined to say that that person does not think nor reason either.

Then were there many coral-reefs in Britain in old times?

Yes, many and many, again and again; some whole ages older than this, a bit of which you see, and some again whole ages newer. But look: then judge for yourself. Look at this geological map. Wherever you see a bit of blue, which is the mark for limestone, you may say, "There is a bit of old coral-reef rising up to the surface." But because I will not puzzle your little head with too many things at once, you shall look at one set of coral-reefs which are far newer than this bit of Dudley limestone, and which are the largest, I suppose, that ever were in this country; or, at least, there is more of them left than of any others.

Look first at Ireland. You see that almost all the middle of Ireland is coloured blue. It is one great sheet of old coral-reef and coral-mud, which is now called the carboniferous limestone. You see red and purple patches rising out of it, like islands—and islands I suppose they were, of hard and ancient rock, standing up in the middle of the coral sea.

But look again, and you will see that along the west coast of Ireland, except in a very few places, like Galway Bay, the blue limestone does not come down to the sea; the shore is coloured purple and brown, and those colours mark the ancient rocks and high mountains of Mayo and Galway and Kerry, which stand as barriers to keep the raging surf of the Atlantic from bursting inland and beating away, as it surely would in course of time, the low flat limestone plain of the middle of Ireland. But the same coral-reefs once stretched out far to the westward into the Atlantic Ocean; and you may see the proof upon that map. For in the western bays, in Clew Bay with its hundred islands, and Galway Bay with its Isles of Arran, and beautiful

Kenmare, and beautiful Bantry, you see little blue spots, which are low limestone islands, standing in the sea, overhung by mountains far aloft. You have often heard those islands in Kenmare Bay talked of, and how some whom you know go to fish round them by night for turbot and conger; and when you hear them spoken of again, you must recollect that they are the last fragments of a great fringing coral-reef, which will in a few thousand years follow the fate of the rest, and be eaten up by the waves, while the mountains of hard rock stand round them still unchanged.

Now look at England, and there you will see patches at least of a great coral-reef which was forming at the same time as that Irish one, and on which perhaps some of your schoolfellows have often stood. You have heard of St. Vincent's Rocks at Bristol, and the marble cliffs, 250 feet in height, covered in part with rich wood and rare flowers, and the Avon running through the narrow gorge, and the stately ships sailing far below your feet from Bristol to the Severn sea. And you may see, for here they are, corals from St. Vincent's Rocks, cut and polished, showing too that they also, like the Dudley limestone, are made up of corals and of coral-mud. Now, whenever you see St. Vincent's Rocks, as I suspect you very soon will, recollect where you are, and use your fancy, to paint for yourself a picture as strange as it is true. Fancy that those rocks are what they once were, a coral-reef close to the surface of a shallow sea. Fancy that there is no gorge of the Avon, no wide Severn sea—for those were eaten out by water ages and ages afterwards. But picture to yourself the coral sea reaching away to the north, to the foot of the Welsh mountains; and then fancy yourself, if you will, in a canoe, paddling up through the coral-reefs, north and still north, up the valley down which the Severn now flows, up through what is now Worcestershire, then up through Staffordshire, then through Derbyshire, into Yorkshire, and so on through Durham and Northumberland, till your find yourself stopped by the Ettrick hills in Scotland; while all to the westward of you, where is now the greater part of England, was open sea. You may say, if you know anything of the geography of England, "Impossible! That would be to paddle over the tops of high mountains; over the top of the Peak in Derbyshire, over the top of High Craven and Whernside and Pen-y-gent and Cross Fell, and to paddle too over the Cheviot Hills, which part England and Scotland." I know it, my child, I know it. But so it was once on a time. The high limestone mountains which part Lancashire and Yorkshire— the very chine and backbone of England—were once coral-reefs at the bottom of the sea. They are all made up of the carboniferous limestone, so called, as your little knowledge of Latin ought to tell you, because it carries the coal; because the coalfields usually lie upon it. It may be impossible in your eyes: but remember always that nothing is impossible with God.

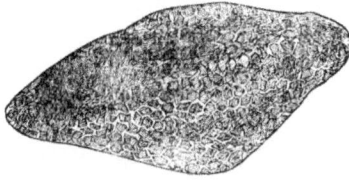

But you said that the coal was made from plants and trees, and did plants and trees grow on this coral-reef?

That I cannot say. Trees may have grown on the dry parts of the reef, as cocoa-nuts grow now in the Pacific. But the coal was not laid down upon it till long afterwards, when it had gone through many and strange changes. For all through the chine of England, and in a part of Ireland too, there lies upon the top of the limestone a hard gritty rock, in some places three thousand feet thick, which is commonly called "the mill-stone grit." And above that again the coal begins. Now to make that 3000 feet of hard rock, what must have happened? The sea-bottom must have sunk, slowly no doubt, carrying the coral-reefs down with it, 3000 feet at least. And meanwhile sand and mud, made from the wearing away of the old lands in the North must have settled down upon it. I say from the North—for there are no fossils, as far as I know, or sign of life, in these rocks of mill-stone grit; and therefore it is reasonable to suppose that they were brought from a cold current at the Pole, too cold to allow sea-beasts to live,—quite cold enough, certainly, to kill coral insects, who could only thrive in warm water coming from the South.

Then, to go on with my story, upon the top of these mill-stone grits came sand and mud, and peat, and trees, and plants, washed out to sea, as far as we can guess, from the mouths of vast rivers flowing from the West, rivers as vast as the Amazon, the Mississippi, or the Orinoco are now; and so in long ages, upon the top of the limestone and upon the top of the mill-stone grit, were laid down those beds of coal which you see burnt now in every fire.

But how did the coral-reefs rise till they became cliffs at Bristol and mountains in Yorkshire?

The earthquake steam, I suppose, raised them. One earthquake indeed, or series of earthquakes, there was, running along between Lancashire and Yorkshire, which made that vast crack and upheaval in the rocks, the Craven Fault, running, I believe, for more than a hundred miles, and lifting the rocks in some places several hundred feet. That earthquake helped to make the high hills which overhang Manchester and Preston, and all the manufacturing county of Lancashire. That earthquake helped to make the perpendicular cliff at Malham Cove, and many another beautiful bit of scenery. And that and other earthquakes, by heating the rocks from the fires below, may have helped to change them from soft coral into hard crystalline marble as you see them now, just as volcanic heat has hardened and purified the beautiful white marbles of Pentelicus and Paros in Greece, and Carrara in Italy, from which statues are carved unto this day. Or the same earthquake may have heated and hardened the limestones simply by grinding and squeezing them; or they may have been heated and hardened in the course of long ages simply by the weight of the thousands of feet of other rock which lay upon them. For pressure, you must remember, produces heat. When you strike flint and steel together, the pressure of the blow not only makes bits of steel fly off, but makes them fly off in red-hot sparks. When you hammer a piece of iron with a hammer, you will soon find it get quite warm. When you squeeze the air together in your pop-gun, you actually make the air inside warmer, till the pellet flies out, and the air expands and cools again. Nay, I believe you cannot hold up a stone on the palm of your hand without that stone after a while warming your hand, because it presses against you in trying to fall, and you press against it in trying to hold it up. And recollect above all the great and beautiful example of that law which you were lucky enough to see on the night of the 14th of November 1867, how those falling stars, as I told you then, were coming out of boundless space, colder than any ice on earth, and yet, simply by pressing against the air above our heads, they had their motion turned into heat, till they burned themselves up into trains of fiery dust. So remember that wherever you have pressure you have heat, and that the pressure of the upper rocks upon the lower is quite enough, some think, to account for the older and lower rocks being harder than the upper and newer ones.

But why should the lower rocks be older and the upper rocks newer? You told me just now that the high mountains in Wales were ages older than Windsor Forest, upon which we stand: but yet how much lower we are here than if we were on a Welsh mountain.

Ah, my dear child, of course that puzzles you, and I am afraid it must puzzle you still till we have another talk; or rather it seems to me that the best way to explain that puzzle to you would be for you and me to go a journey into the far west, and look into the matter for ourselves; and from here to the far west we will go, either in fancy or on a real railroad and steamboat, before we have another talk about these things.

Now it is time to stop. Is there anything more you want to know? for you look as if something was puzzling you still.

Were there any men in the world while all this was going on?

I think not. We have no proof that there were not: but also we have no proof that there were; the cave-men, of whom I told you, lived many ages after the coal was covered up. You seem to be sorry that there were no men in the world then.

Because it seems a pity that there was no one to see those beautiful coral-reefs and coal-forests.

No one to see them, my child? Who told you that? Who told you there are not, and never have been any rational beings in this vast universe, save certain weak, ignorant, short-sighted creatures shaped like you and me? But even if it were so, and no created eye had ever beheld those ancient wonders, and no created heart ever enjoyed them, is there not one Uncreated who has seen them and enjoyed them from the beginning? Were not these creatures enjoying themselves each after their kind? And was there not a Father in Heaven who was enjoying their enjoyment, and enjoying too their beauty, which He had formed according to the ideas of His Eternal Mind? Recollect what you were told on Trinity Sunday—That this world was not made for man alone: but that man, and this world, and the whole Universe was made for God; for He created all things, and for His pleasure they are, and were created.

Chapter X. Field and Wild

Where were we to go next? Into the far west, to see how all the way along the railroads the new rocks and soils lie above the older, and yet how, when we get westward, the oldest rocks rise highest into the air.

Well, we will go: but not, I think, to-day. Indeed I hardly know how we could get as far as Reading; for all the world is in the hay-field, and even the old horse must go thither too, and take his turn at the hay-cart. Well, the rocks have been where they are for many a year, and they will wait our leisure patiently enough: but Midsummer and the hay-field will not wait. Let us take what God gives when He sends it, and learn the lesson that lies nearest to us. After all, it is more to my old mind, and perhaps to your young mind too, to look at things which are young and fresh and living, rather than things which are old and worn and dead. Let us leave the old stones, and the old bones, and the old shells, the wrecks of ancient worlds which have gone down into the kingdom of death, to teach us their grand lessons some other day; and let us look now at the world of light and life and beauty, which begins here at the open door, and stretches away over the hay-fields, over the woods, over the southern moors, over sunny France, and sunnier Spain, and over the tropic seas, down to the equator, and the palm-groves of the eternal summer. If we cannot find something, even at starting from the open door, to teach us about Why and How, we must be very short-sighted, or very shallow-hearted.

THE STARLING

There is the old cock starling screeching in the eaves, because he wants to frighten us away, and take a worm to his children, without our finding out whereabouts his hole is. How does he know that we might hurt him? and how again does he not know that we shall not hurt him? we, who for five-and-twenty years have let him and his ancestors build under those eaves in peace? How did he get that quantity of half-wit, that sort of stupid cunning, into his little brain, and yet get no more? And why (for this is a question of Why, and not of How) does he labour all day long, hunting for worms and insects for his children, while his wife nurses them in the nest? Why, too, did he help her to build that nest with toil and care this spring, for the sake of a set of nestlings who can be of no gain or use to him, but only take the food out of his mouth? Simply out of—what shall I call it, my child?—Love; that same sense of love and duty, coming surely from that one Fountain of all duty and all love, which makes your father work for you. That the mother should take care of her young, is wonderful enough; but that (at least among many birds) the father should help likewise, is (as you will find out as you grow older) more wonderful far. So there already the old starling has set us two fresh puzzles about How and Why, neither of which we shall get answered, at least on this side of the grave.

Come on, up the field, under the great generous sun, who quarrels with no one, grudges no one, but shines alike upon the evil and the good. What a gay picture he is painting now, with his light-pencils; for in them, remember, and not in the things themselves the colour lies. See how, where the hay has been already carried, he floods all the slopes with yellow light, making them stand out sharp against the black shadows of the wood; while where the grass is standing still, he makes the sheets of sorrel-flower blush rosy red, or dapples the field with white oxeyes.

But is not the sorrel itself red, and the oxeyes white?

What colour are they at night, when the sun is gone?

Dark.

That is, no colour. The very grass is not green at night.

Oh, but it is if you look at it with a lantern.

No, no. It is the light of the lantern, which happens to be strong enough to make the leaves look green, though it is not strong enough to make a geranium look red.

Not red?

No; the geranium flowers by a lantern look black, while the leaves look green. If you don't believe me, we will try.

But why is that?

Why, I cannot tell: and how, you had best ask Professor Tyndall, if you ever have the honour of meeting him.

But now—hark to the mowing-machine, humming like a giant night-jar. Come up and look at it, and see how swift and smooth it shears the long grass down, so that in the middle of the swathe it seems to have merely fallen flat, and you must move it before you find that it has been cut off.

Ah, there is a proof to us of what men may do if they will only learn the lessons which Madam How can teach them. There is that boy, fresh from the National School, cutting more grass in a day than six strong mowers could have cut, and cutting it better, too; for the mowing-machine goes so much nearer to the ground than the scythe, that we gain by it two hundredweight of hay on every acre. And see, too, how persevering old Madam How will not stop her work, though the machine has cut off all the grass which she has been making for the last three months; for as fast as we shear it off, she makes it grow again. There are fresh blades, here at our feet, a full inch long, which have sprung up in the last two days, for the cattle when they are turned in next week.

But if the machine cuts all the grass, the poor mowers will have nothing to do.

Not so. They are all busy enough elsewhere. There is plenty of other work to be done, thank God; and wholesomer and easier work than mowing with a burning sun on their backs, drinking gallons of beer, and getting first hot and then cold across the loins, till they lay in a store of lumbago and sciatica, to cripple them in their old age. You delight in machinery because it is curious: you should delight in it besides because it does good, and nothing but good, where it is used, according to the laws of Lady Why, with care, moderation, and mercy, and fair-play between man and man. For example: just as the mowing-machine saves the mowers, the threshing-machine saves the threshers from rheumatism and chest complaints,—which they used to catch in the draught and dust of the unhealthiest place in the whole parish, which is, the old-fashioned barn's floor. And so, we may hope, in future years all heavy drudgery and dirty work will be done more and more by machines, and people will have more and more chance of keeping themselves clean and healthy, and more and more time to read, and learn, and think, and be true civilized men and women, instead of being mere live ploughs, or live manure-carts, such as I have seen ere now.

A live manure-cart?

Yes, child. If you had seen, as I have seen, in foreign lands, poor women, haggard, dirty, grown old before their youth was over, toiling up hill with baskets of foul manure upon their backs, you would have said, as I have said, "Oh for Madam How to cure that ignorance! Oh for Lady Why to cure that barbarism! Oh that Madam How would teach them that machinery must always be cheaper in the long run than human muscles and nerves! Oh that Lady Why would teach them that a woman is the most precious thing on earth, and that if she be turned into a beast of burden, Lady Why—and Madam How likewise—will surely avenge the wrongs of their human sister!" There, you do not quite know what I mean, and I do not care that you should. It is good for little folk that big folk should now and then "talk over their heads," as the saying is, and make them feel how ignorant they are, and how many solemn and earnest questions there are in the world on which they must make up their minds some day, though not yet. But now we will talk about the hay: or rather do you and the rest go and play in the hay and gather it up, build forts of it, storm them, pull them down, build them up again, shout, laugh, and scream till you are hot and tired.

You will please Madam How thereby, and Lady Why likewise.

How?

Because Madam How naturally wants her work to succeed, and she is at work now making you.

Making me?

Of course. Making a man of you, out of a boy. And that can only be done by the life-blood which runs through and through you. And the more you laugh and shout, the more pure air will pass into your blood, and make it red and healthy; and the more you romp and play—unless you overtire yourself—the quicker will that blood flow through all your limbs, to make bone and muscle, and help you to grow into a man.

But why does Lady Why like to see us play?

She likes to see you happy, as she likes to see the trees and birds happy. For she knows well that there is no food, nor medicine either, like happiness. If people are not happy enough, they are often tempted to do many wrong deeds, and to think many wrong thoughts: and if by God's grace they know the laws of Lady Why, and keep from sin, still unhappiness, if it goes on too long, wears them out, body and mind; and they grow ill and die, of broken hearts, and broken brains, my child; and so at last, poor souls, find "Rest beneath the Cross."

Children, too, who are unhappy; children who are bullied, and frightened, and kept dull and silent, never thrive. Their bodies do not thrive; for they grow up weak. Their minds do not thrive; for they grow up dull. Their souls do not thrive; for they learn mean, sly, slavish ways, which God forbid you should ever learn. Well said the wise man, "The human plant, like the vegetables, can only flower in sunshine."

So do you go, and enjoy yourself in the sunshine; but remember this—You know what happiness is. Then if you wish to please Lady Why, and Lady Why's Lord and King likewise, you will never pass a little child without trying to make it happier, even by a passing smile. And now be off, and play in the hay, and come back to me when you are tired.

* * * * *

Let us lie down at the foot of this old oak, and see what we can see.

And hear what we can hear, too. What is that humming all round us, now that the noisy mowing-machine has stopped?

And as much softer than the noise of mowing-machine hum, as the machines which make it are more delicate and more curious. Madam How is a very skilful workwoman, and has eyes which see deeper and clearer than all microscopes; as you would find, if you tried to see what makes that "Midsummer hum" of which the haymakers are so fond, because it promises fair weather.

Why, it is only the gnats and flies.

Only the gnats and flies? You might study those gnats and flies for your whole life without finding out all—or more than a very little—about them. I wish I knew how they move those tiny wings of theirs—a thousand times in a second, I dare say, some of them. I wish I knew how far they know that they are happy—for happy they must be, whether they know it or not. I wish I knew how they live at all. I wish I even knew how many sorts there are humming round us at this moment.

How many kinds? Three or four?

More probably thirty or forty round this single tree.

But why should there be so many kinds of living things? Would not one or two have done just as well?

Why, indeed? Why should there not have been only one sort of butterfly, and he only of one colour, a plain brown, or a plain white?

And why should there be so many sorts of birds, all robbing the garden at once? Thrushes, and blackbirds, and sparrows, and chaffinches, and greenfinches, and bullfinches, and tomtits.

And there are four kinds of tomtits round here, remember: but we may go on with such talk for ever. Wiser men than we have asked the same question: but Lady Why will not answer them yet. However, there is another question, which Madam How seems inclined to answer just now, which is almost as deep and mysterious.

What?

How all these different kinds of things became different.

Oh, do tell me!

Not I. You must begin at the beginning, before you can end at the end, or even make one step towards the end.

What do you mean?

You must learn the differences between things, before you can find out how those differences came about. You must learn Madam How's alphabet before you can read her book. And Madam How's alphabet of animals and plants is, Species, Kinds of things. You must see which are like, and which unlike; what they are like in, and what they are unlike in. You are beginning to do that with your collection of butterflies. You like to arrange them, and those that are most like nearest to each other, and to compare them. You must do that with thousands of different kinds of things before you can read one page of Madam How's Natural History Book rightly.

But it will take so much time and so much trouble.

God grant that you may not spend more time on worse matters, and take more trouble over things which will profit you far less. But so it must be, willy-nilly. You must learn the alphabet if you mean to read. And you must learn the value of the figures before you can do a sum. Why, what would you think of any one who sat down to play at cards—for money too (which I hope and trust you never will do)—

before he knew the names of the cards, and which counted highest, and took the other?

Of course he would be very foolish.

Just as foolish are those who make up "theories" (as they call them) about this world, and how it was made, before they have found out what the world is made of. You might as well try to find out how this hay-field was made, without finding out first what the hay is made of.

How the hay-field was made? Was it not always a hay-field?

Ah, yes; the old story, my child: Was not the earth always just what it is now? Let us see for ourselves whether this was always a hay-field.

How?

Just pick out all the different kinds of plants and flowers you can find round us here. How many do you think there are?

MOUSE-EAR HACKWEED

Oh—there seem to be four or five.

Just as there were three or four kinds of flies in the air. Pick them, child, and count. Let us have facts.

How many? What! a dozen already?

Yes—and here is another, and another. Why, I have got I don't know how many.

Why not? Bring them here, and let us see. Nine kinds of grasses, and a rush. Six kinds of clovers and vetches; and besides, dandelion, and rattle, and oxeye, and sorrel, and plantain, and buttercup, and a little stitchwort, and pignut, and mouse-ear hawkweed, too, which nobody wants.

Why?

Because they are a sign that I am not a good farmer enough, and have not quite turned my Wild into Field.

What do you mean?

Look outside the boundary fence, at the moors and woods; they are forest, Wild—"Wald," as the Germans would call it. Inside the fence is Field—"Feld," as the Germans would call it. Guess why?

Is it because the trees inside have been felled?

Well, some say so, who know more than I. But now go over the fence, and see how many of these plants you can find on the moor.

Oh, I think I know. I am so often on the moor.

I think you would find more kinds outside than you fancy. But what do you know?

That beside some short fine grass about the cattle-paths, there are hardly any grasses on the moor save deer's hair and glade-grass; and all the rest is heath, and moss, and furze, and fern.

Softly—not all; you have forgotten the bog plants; and there are (as I said) many more plants beside on the moor than you fancy. But we will look into that another time. At all events, the plants outside are on the whole quite different from the hay-field.

Of course: that is what makes the field look green and the moor brown.

Not a doubt. They are so different, that they look like bits of two different continents. Scrambling over the fence is like scrambling out of Europe into Australia. Now, how was that difference made? Think. Don't guess, but think. Why does the rich grass come up to the bank, and yet not spread beyond it?

I suppose because it cannot get over.

Not get over? Would not the wind blow the seeds, and the birds carry them? They do get over, in millions, I don't doubt, every summer.

Then why do they not grow?

Think.

Is there any difference in the soil inside and out?

A very good guess. But guesses are no use without facts. Look.

Oh, I remember now. I know now the soil of the field is brown, like the garden; and the soil of the moor all black and peaty.

Yes. But if you dig down two or three feet, you will find the soils of the moor and the field just the same. So perhaps the top soils were once both alike.

I know.

Well, and what do you think about it now? I want you to look and think. I want every one to look and think. Half the misery in the world comes first from not looking, and then from not thinking. And I do not want you to be miserable.

But shall I be miserable if I do not find out such little things as this.

You will be miserable if you do not learn to understand little things: because then you will not be able to understand great things when you meet them. Children who are not trained to use their eyes and their common sense grow up the more miserable the cleverer they are.

Why?

Because they grow up what men call dreamers, and bigots, and fanatics, causing misery to themselves and to all who deal with them. So I say again, think.

Well, I suppose men must have altered the soil inside the bank.

Well done. But why do you think so?

Because, of course, some one made the bank; and the brown soil only goes up to it.

Well, that is something like common sense. Now you will not say any more, as the cows or the butterflies might, that the hay-field was always there.

And how did men change the soil?

By tilling it with the plough, to sweeten it, and manuring it, to make it rich.

And then did all these beautiful grasses grow up of themselves?

You ought to know that they most likely did not. You know the new enclosures?

Yes.

Well then, do rich grasses come up on them, now that they are broken up?

Oh no, nothing but groundsel, and a few weeds.

Just what, I dare say, came up here at first. But this land was tilled for corn, for hundreds of years, I believe. And just about one hundred years ago it was laid down in grass; that is, sown with grass seeds.

And where did men get the grass seeds from?

Ah, that is a long story; and one that shows our forefathers (though they knew nothing about railroads or electricity) were not such simpletons as some folks think. The way it must have been done was this. Men watched the natural pastures where cattle get fat on the wild grass, as they do in the Fens, and many other parts of England. And then they saved the seeds of those fattening wild grasses, and sowed them in fresh spots. Often they made mistakes. They were careless, and got weeds among the seed—like the buttercups, which do so much harm to this pasture. Or they sowed on soil which would not suit the seed, and it died. But at last, after many failures, they have grown so careful and so clever, that you may send to certain shops, saying what sort of soil yours is, and they will send you just the seeds which will grow there, and no other; and then you have a good pasture for as long as you choose to keep it good.

And how is it kept good?

Look at all those loads of hay, which are being carried off the field. Do you think you can take all that away without putting anything in its place?

Why not?

If I took all the butter out of the churn, what must I do if I want more butter still?

Put more cream in.

So, if I want more grass to grow, I must put on the soil more of what grass is made of.

But the butter don't grow, and the grass does.

What does the grass grow in?

The soil.

Yes. Just as the butter grows in the churn. So you must put fresh grass-stuff continually into the soil, as you put fresh cream into the churn. You have heard the farm men say, "That crop has taken a good deal out of the land"?

Yes.

Then they spoke exact truth. What will that hay turn into by Christmas? Can't you tell? Into milk, of course, which you will drink; and into horseflesh too, which you will use.

Use horseflesh? Not eat it?

No; we have not got as far as that. We did not even make up our minds to taste the Cambridge donkey. But every time the horse draws the carriage, he uses up so much muscle; and that muscle he must get back again by eating hay and corn; and that hay and corn must be put back again into the land by manure, or there will be all the less for the horse next year. For one cannot eat one's cake and keep it too; and no more can one eat one's grass.

So this field is a truly wonderful place. It is no ugly pile of brick and mortar, with a tall chimney pouring out smoke and evil smells, with unhealthy, haggard people toiling inside. Why do you look surprised?

Because—because nobody ever said it was. You mean a manufactory.

Well, and this hay-field is a manufactory: only like most of Madam How's workshops, infinitely more beautiful, as well as infinitely more crafty, than any manufactory of man's building. It is beautiful to behold, and healthy to work in; a joy and blessing alike to the eye, and the mind, and the body: and yet it is a manufactory.

But a manufactory of what?

Of milk of course, and cows, and sheep, and horses; and of your body and mine—for we shall drink the milk and eat the meat. And therefore it is a flesh and milk manufactory. We must put into it every year yard-stuff, tank-stuff, guano, bones, and anything and everything of that kin, that Madam How may cook it for us into grass, and cook the grass again into milk and meat. But if we don't give Madam How material to work on, we cannot expect her to work for us. And what do you think will happen then? She will set to work for herself. The rich grasses will dwindle for want of ammonia (that is smelling salts), and the rich clovers for want of phosphates (that is bone-earth): and in their places will come over the bank the old weeds and grass off the moor, which have not room to get in now, because the ground is coveted already. They want no ammonia nor phosphates—at all events they have none, and that is why the cattle on the moor never get fat. So they can live where these rich grasses cannot. And then they will conquer and thrive; and the Field will turn into Wild once more.

Ah, my child, thank God for your forefathers, when you look over that boundary mark. For the difference between the Field and the Wild is the difference between the old England of Madam How's making, and the new England which she has taught man to make, carrying on what she had only begun and had not time to finish.

That moor is a pattern bit left to show what the greater part of this land was like for long ages after it had risen out of the sea; when there was little or nothing on the flat upper moors save heaths, and ling, and club-mosses, and soft gorse, and needle-whin, and creeping willows; and furze and fern upon the brows; and in the bottoms oak and ash, beech and alder, hazel and mountain ash, holly and thorn, with here and there an aspen or a buckthorn (berry-bearing alder as you call it), and everywhere—where he could thrust down his long root, and thrust up his long shoots—that intruding conqueror and insolent tyrant, the bramble. There were sedges and rushes, too, in the bogs, and coarse grass on the forest pastures—or "leas" as we call them to this day round here—but no real green fields; and, I suspect, very few gay flowers, save in spring the sheets of golden gorse, and in summer the purple heather. Such was old England—or rather, such was this land before it was England; a far sadder, damper, poorer land than now. For one man or one cow or sheep which could have lived on it then, a hundred can live now. And yet, what it was once, that it might become again,—it surely would round here, if this brave English people died out of it, and the land was left to itself once more.

What would happen then, you may guess for yourself, from what you see happen whenever the land is left to itself, as it is in the wood above. In that wood you can still see the grass ridges and furrows which show that it was once ploughed and sown by man; perhaps as late as the time of Henry the Eighth, when a great deal of poor land, as you will read some day, was thrown out of tillage, to become forest and down once more. And what is the mount now? A jungle of oak and beech, cherry and holly, young and old all growing up together, with the mountain ash and bramble and furze coming up so fast beneath them, that we have to cut the paths clear again year by year. Why, even the little cow-wheat, a very old-world plant, which only grows in ancient woods, has found its way back again, I know not whence, and covers the open spaces with its pretty yellow and white flowers. Man had conquered this mount, you see, from Madam How, hundreds of years ago. And she always lets man conquer her, because Lady Why wishes man to conquer: only he must have a fair fight with Madam How first, and try his strength against hers to the utmost. So man conquered the wood for a while; and it became cornfield instead of forest: but he was not strong and wise enough three hundred years ago to keep what he had conquered; and back came Madam How, and took the place into her own hands, and bade the old forest trees and plants come back again—as they would come if they were not stopped year by year, down from the wood, over the pastures—killing the rich grasses as they went, till they met another forest coming up from below, and fought it for many a year, till both made peace, and lived quietly side by side for ages.

COW-WHEAT

Another forest coming up from below? Where would it come from?

From where it is now. Come down and look along the brook, and every drain and grip which runs into the brook. What is here?

Seedling alders, and some withies among them.

Very well. You know how we pull these alders up, and cut them down, and yet they continually come again. Now, if we and all human beings were to leave this pasture for a few hundred years, would not those alders increase into a wood? Would they not kill the grass, and spread right and left, seeding themselves more and more as the grass died, and left the ground bare, till they met the oaks and beeches coming down the hill? And then would begin a great fight, for years and years, between oak and beech against alder and willow.

But how can trees fight? Could they move or beat each other with their boughs?

Not quite that; though they do beat each other with their boughs, fiercely enough, in a gale of wind; and then the trees who have strong and stiff boughs wound those who have brittle and limp boughs, and so hurt them, and if the storms come often enough, kill them. But among these trees in a sheltered valley the larger and stronger would kill the weaker and smaller by simply overshadowing their tops, and starving their roots; starving them, indeed, so much when they grow very thick, that the poor little acorns, and beech mast, and alder seeds would not be able to sprout at all. So they would fight, killing each other's children, till the war ended—I think I can guess how.

How?

The beeches are as dainty as they are beautiful; and they do not like to get their feet wet. So they would venture down the hill only as far as the dry ground lasts, and those who tried to grow any lower would die. But the oaks are hardy, and do not care much where they grow. So they would fight their way down into the wet ground among the alders and willows, till they came to where their enemies were so thick and tall, that the acorns as they fell could not sprout in the darkness. And so you would have at last, along the hill-side, a forest of beech and oak, lower down a forest of oak and alder, and along the stream-side alders and willows only. And that would be a very fair example of the great law of the struggle for existence, which causes the competition of species.

What is that?

Madam How is very stern, though she is always perfectly just; and therefore she makes every living thing fight for its life, and earn its bread, from its birth till its death; and rewards it exactly according to its deserts, and neither more nor less.

And the competition of species means, that each thing, and kind of things, has to compete against the things round it; and to see which is the stronger; and the stronger live, and breed, and spread, and the weaker die out.

But that is very hard.

I know it, my child, I know it. But so it is. And Madam How, no doubt, would be often very clumsy and very cruel, without meaning it, because she never sees beyond her own nose, or thinks at all about the consequences of what she is doing. But Lady Why, who does think about consequences, is her mistress, and orders her about for ever. And Lady Why is, I believe, as loving as she is wise; and therefore we must trust that she guides this great war between living things, and takes care that Madam How kills nothing which ought not to die, and takes nothing away without putting something more beautiful and something more useful in its place; and that even if England were, which God forbid, overrun once more with forests and bramble-brakes, that too would be of use somehow, somewhere, somewhen, in the long ages which are to come hereafter.

And you must remember, too, that since men came into the world with rational heads on their shoulders, Lady Why has been handing over more and more of Madam How's work to them, and some of her own work too: and bids them to put beautiful and useful things in the place of ugly and useless ones; so that now it is men's own fault if they do not use their wits, and do by all the world what they have done by these pastures—change it from a barren moor into a rich hay-field, by copying the laws of Madam How, and making grass compete against heath. But you look thoughtful: what is it you want to know?

Why, you say all living things must fight and scramble for what they can get from each other: and must not I too? For I am a living thing.

Ah, that is the old question, which our Lord answered long ago, and said, "Be not anxious what ye shall eat or what ye shall drink, or wherewithal you shall be clothed. For after all these things do the heathen seek, and your Heavenly Father knoweth that ye have need of these things. But seek ye first the kingdom of God and His righteousness, and all these things shall be added to you." A few, very few, people have taken that advice. But they have been just the salt of the earth, which has kept mankind from decaying.

But what has that to do with it?

See. You are a living thing, you say. Are you a plant?

No.

Are you an animal?

I do not know. Yes. I suppose I am. I eat, and drink, and sleep, just as dogs and cats do.

Yes. There is no denying that. No one knew that better than St. Paul when he told men that they had a flesh; that is, a body, and an animal's nature in them. But St. Paul told them—of course he was not the first to say so, for all the wise heathens have known that—that there was something more in us, which he called a spirit. Some call it now the moral sentiment, some one thing, some another, but we will keep to the old word: we shall not find a better.

Yes, I know that I have a spirit, a soul.

Better to say that you are a spirit. But what does St. Paul say? That our spirit is to conquer our flesh, and keep it down. That the man in us, in short, which is made in the likeness of God, is to conquer the animal in us, which is made in the likeness of the dog and the cat, and sometimes (I fear) in the likeness of the ape or the pig. You would not wish to be like a cat, much less like an ape or a pig?

Of course not.

Then do not copy them, by competing and struggling for existence against other people.

What do you mean?

Did you never watch the pigs feeding?

Yes, and how they grudge and quarrel, and shove each other's noses out of the trough, and even bite each other because they are so jealous which shall get most.

That is it. And how the biggest pig drives the others away, and would starve them while he got fat, if the man did not drive him off in his turn.

Oh, yes; I know.

Then no wiser than those pigs are worldly men who compete, and grudge, and struggle with each other, which shall get most money, most fame, most power over their fellow-men. They will tell you, my child, that that is the true philosophy, and the true wisdom; that competition is the natural law of society, and the source of wealth and prosperity. Do not you listen to them. That is the wisdom of this world, which the flesh teaches the animals; and those who follow it, like the animals, will perish. Such men are not even as wise as Sweep the retriever.

Not as wise as Sweep?

Not they. Sweep will not take away Victor's bone, though he is ten times as big as Victor, and could kill him in a moment; and when he catches a rabbit, does he eat it himself?

Of course not; he brings it and lays it down at our feet.

Because he likes better to do his duty, and be praised for it, than to eat the rabbit, dearly as he longs to eat it.

But he is only an animal. Who taught him to be generous, and dutiful, and faithful?

Who, indeed! Not we, you know that, for he has grown up with us since a puppy. How he learnt it, and his parents before him, is a mystery, of which we can only say, God has taught them, we know not how. But see what has happened—that just because dogs have learnt not to be selfish and to compete—that is, have become civilised and tame—therefore we let them live with us, and love them. Because they try to be good in their simple way, therefore they too have all things added to them, and live far happier, and more comfortable lives than the selfish wolf and fox.

But why have not all animals found out that?

I cannot tell: there may be wise animals and foolish animals, as there are wise and foolish men. Indeed there are. I see a very wise animal there, who never competes; for she has learned something of the golden lesson—that it is more blessed to give than to receive; and she acts on what she has learnt, all day long.

Which do you mean? Why, that is a bee.

Yes, it is a bee: and I wish I were as worthy in my place as that bee is in hers. I wish I could act up as well as she does to the true wisdom, which is self-sacrifice. For whom is that bee working? For herself? If that was all, she only needs to suck the honey as she goes. But she is storing up the wax under her stomach, and bee-bread in her thighs—for whom? Not for herself only, or even for her own children: but for the children of another bee, her queen. For them she labours all day long, builds for them, feeds them, nurses them, spends her love and cunning on them. So does that ant on the path. She is carrying home that stick to build for other ants' children. So do the white ants in the tropics. They have learnt not to compete, but to help each other; not to be selfish, but to sacrifice themselves; and therefore they are strong.

But you told me once that ants would fight and plunder each other's nests. And once we saw two hives of bees fighting in the air, and falling dead by dozens.

My child, do not men fight, and kill each other by thousands with sharp shot and cold steel, because, though they have learnt the virtue of patriotism, they have not yet learnt that of humanity? We must not blame the bees and ants if they are no wiser than men. At least they are wise enough to stand up for their country, that is, their hive, and work for it, and die for it, if need be; and that makes them strong.

But how does that make them strong?

How, is a deep question, and one I can hardly answer yet. But that it has made them so there is no doubt. Look at the solitary bees—the governors as we call them, who live in pairs, in little holes in the banks. How few of them there are; and they never seem to increase in numbers. Then look at the hive bees, how, just because they are civilized,—that is, because they help each other, and feed each other, instead of being solitary and selfish,—they breed so fast, and get so much food, that if they were not killed for their honey, they would soon become a nuisance, and drive us out of the parish.

But then we give them their hives ready made.

True. But in old forest countries, where trees decay and grow hollow, the bees breed in them.

Yes. I remember the bee tree in the fir avenue.

Well then, in many forests in hot countries the bees swarm in hollow trees; and they, and the ants, and the white ants, have it all their own way, and are lords and masters, driving the very wild beasts before them, while the ants and white ants eat up all gardens, and plantations, and clothes, and furniture; till it is a serious question whether in some hot countries man will ever be able to settle, so strong have the ants grown, by ages of civilization, and not competing against their brothers and sisters.

But may I not compete for prizes against the other boys?

Well, there is no harm in that; for you do not harm the others, even if you win. They will have learnt all the more, while trying for the prize; and so will you, even if you don't get it. But I tell you fairly, trying for prizes is only fit for a child; and when you become a man, you must put away childish things—competition among the rest.

But surely I may try to be better and wiser and more learned than everybody else?

My dearest child, why try for that? Try to be as good, and wise, and learned as you can, and if you find any man, or ten thousand men, superior to you, thank God for it. Do you think that there can be too much wisdom in the world?

Of course not: but I should like to be the wisest man in it.

Then you would only have the heaviest burden of all men on your shoulders.

Why?

Because you would be responsible for more foolish people than any one else. Remember what wise old Moses said, when some one came and told him that certain men in the camp were prophesying—"Would God all the Lord's people did prophesy!" Yes; it would have saved Moses many a heartache, and many a sleepless night, if all the Jews had been wise as he was, and wiser still. So do not you compete with good and wise men, but simply copy them: and whatever you do, do not compete with the wolves, and the apes, and the swine of this world; for that is a game at which you are sure to be beaten.

Why?

Because Lady Why, if she loves you (as I trust she does), will take care that you are beaten, lest you should fancy it was really profitable to live like a cunning sort of animal, and not like a true man. And how she will do that I can tell you. She will take care that you always come across a worse man than you are trying to be,—a more apish man, who can tumble and play monkey-tricks for people's amusement better than you can; or a more swinish man, who can get at more of the pig's-wash than you can; or a more wolfish man, who will eat you up if you do not get out of his way; and so she will disappoint and disgust you, my child, with that greedy, selfish, vain animal life, till you turn round and see your mistake, and try to live the true human life, which also is divine;—to be just and honourable, gentle and forgiving, generous and useful—in one word, to fear God, and keep His commandments: and as you live that life, you will find that, by the eternal laws of Lady Why, all other things will be added to you; that people will be glad to know you, glad to help you, glad to employ you, because they see that you will be of use to them, and will do them no harm. And if you meet (as you will meet) with people better and wiser than yourself, then so much the better for you; for they will love you, and be glad to teach you when they see that you are living the unselfish and harmless life; and that you come to them, not as foolish Critias came to Socrates, to learn political cunning, and become a selfish and ambitious tyrant, but as wise Plato came, that he might learn the laws of Lady Why, and love them for her sake, and teach them to all mankind. And so you, like the plants and animals, will get your deserts exactly, without competing and struggling for existence as they do.

And all this has come out of looking at the hay-field and the wild moor.

Why not? There is an animal in you, and there is a man in you. If the animal gets the upper hand, all your character will fall back into wild useless moor; if the man gets the upper hand, all your character will be cultivated into rich and fertile field. Choose.

Now come down home. The haymakers are resting under the hedge. The horses are dawdling home to the farm. The sun is getting low, and the shadows long. Come home, and go to bed while the house is fragrant with the smell of hay, and dream that you are still playing among the haycocks. When you grow old, you will have other and sadder dreams.

Chapter XI. The World's End

Hullo! hi! wake up. Jump out of bed, and come to the window, and see where you are.

What a wonderful place!

So it is: though it is only poor old Ireland. Don't you recollect that when we started I told you we were going to Ireland, and through it to the World's End; and here we are now safe at the end of the old world, and beyond us the great Atlantic, and beyond that again, thousands of miles away, the new world, which will be rich and prosperous, civilised and noble, thousands of years hence, when this old world, it may be, will be dead, and little children there will be reading in their history books of Ancient England and of Ancient France, as you now read of Greece and Rome.

But what a wonderful place it is! What are those great green things standing up in the sky, all over purple ribs and bars, with their tops hid in the clouds?

Those are mountains; the bones of some old world, whose poor bare sides Madam How is trying to cover with rich green grass.

And how far off are they?

How I should like to walk up to the top of that one which looks quite close.

You will find it a long walk up there; three miles, I dare say, over black bogs and banks of rock, and up corries and cliffs which you could not climb. There are plenty of cows on that mountain: and yet they look so small, you could not see them, nor I either, without a glass. That long white streak, zigzagging down the mountain side, is a roaring cataract of foam five hundred feet high, full now with last night's rain; but by this afternoon it will have dwindled to a little thread; and to-morrow, when you get up, if no more rain has come down, it will be gone. Madam How works here among the mountains swiftly and hugely, and sometimes terribly enough; as you shall see when you have had your breakfast, and come down to the bridge with me.

But what a beautiful place it is! Flowers and woods and a lawn; and what is that great smooth patch in the lawn just under the window?

Is it an empty flower-bed?

Ah, thereby hangs a strange tale. We will go and look at it after breakfast, and then you shall see with your own eyes one of the wonders which I have been telling you of.

And what is that shining between the trees?

Water.

Is it a lake?

Not a lake, though there are plenty round here; that is salt water, not fresh. Look away to the right, and you see it through the opening of the woods again and again: and now look above the woods. You see a faint blue line, and gray and purple lumps like clouds, which rest upon it far away. That, child, is the great Atlantic Ocean, and those are islands in the far west. The water which washes the bottom of the lawn was but a few months ago pouring out of the Gulf of Mexico, between the Bahamas and Florida, and swept away here as the great ocean river of warm water which we call the Gulf Stream, bringing with it out of the open ocean the shoals of mackerel, and the porpoises and whales which feed upon them. Some fine afternoon we will run down the bay and catch strange fishes, such as you never saw before, and very likely see a living whale.

What? such a whale as they get whalebone from, and which eats sea-moths?

No, they live far north, in the Arctic circle; these are grampuses, and bottle-noses, which feed on fish; not so big as the right whales, but quite big enough to astonish you, if one comes up and blows close to the boat. Get yourself dressed and come down, and then we will go out; we shall have plenty to see and talk of at every step.

Now, you have finished your breakfast at last, so come along, and we shall see what we shall see. First run out across the gravel, and scramble up that bank of lawn, and you will see what you fancied was an empty flower-bed.

Why, it is all hard rock.

Ah, you are come into the land of rocks now: out of the land of sand and gravel; out of a soft young corner of the world into a very hard, old, weather-beaten corner; and you will see rocks enough, and too many for the poor farmers, before you go home again.

But how beautifully smooth and flat the rock is: and yet it is all rounded.

What is it like?

Like—like the half of a shell.

Not badly said, but think again.

Like—like—I know what it is like. Like the back of some great monster peeping up through the turf.

You have got it. Such rocks as these are called in Switzerland "roches moutonnées," because they are, people fancy, like sheep's backs. Now look at the cracks and layers in it. They run across the stone; they have nothing to do with the shape of it. You see that?

Yes: but here are cracks running across them, all along the stone, till the turf hides them.

Look at them again; they are no cracks; they do not go into the stone.

I see. They are scratched; something like those on the elder-stem at home, where the cats sharpen their claws. But it would take a big cat to make them.

Do you recollect what I told you of Madam How's hand, more flexible than any hand of man, and yet strong enough to grind the mountains into paste?

I know. Ice! ice! ice! But are these really ice-marks?

Child, on the place where we now stand, over rich lawns, and warm woods, and shining lochs, lay once on a time hundreds, it may be thousands, of feet of solid ice, crawling off yonder mountain-tops into the ocean there outside; and this is one of its tracks. See how the scratches all point straight down the valley, and straight out to sea. Those mountains are 2000 feet high: but they were much higher once; for the ice has planed the tops off them. Then, it seems to me, the ice sank, and left the mountains standing out of it about half their height, and at that level it stayed, till it had planed down all those lower moors of smooth bare rock between us and the Western ocean; and then it sank again, and dwindled back, leaving moraines (that is, heaps of dirt and stones) all up these valleys here and there, till at the last it melted all away, and poor old Ireland became fit to live in again. We will go down the bay some day and look at those moraines, some of them quite hills of earth, and then you will see for yourself how mighty a chisel the ice-chisel was, and what vast heaps of chips it has left behind. Now then, down over the lawn towards the bridge. Listen to the river, louder and louder every step we take.

What a roar! Is there a waterfall there?

No. It is only the flood. And underneath the roar of that flood, do you not hear a deeper note—a dull rumbling, as if from underground?

Yes. What is it?

The rolling of great stones under water, which are being polished against each other, as they hurry toward the sea. Now, up on the parapet of the bridge. I will hold you tight. Look and see Madam How's rain-spade at work. Look at the terrible yellow torrent below us, almost filling up the arches of the bridge, and leaping high in waves and crests of foam.

Oh, the bridge is falling into the water!

Not a bit. You are not accustomed to see water running below you at ten miles an hour. Never mind that feeling. It will go off in a few seconds. Look; the water is full six feet up the trunks of the trees; over the grass and the king fern, and the tall purple loose-strife—

Oh! Here comes a tree dancing down!

And there are some turfs which have been cut on the mountain. And there is a really sad sight. Look what comes now.

One—two—three.

Why, they are sheep.

Yes. And a sad loss they will be to some poor fellow in the glen above.

And oh! Look at the pig turning round and round solemnly in the corner under the rock. Poor piggy! He ought to have been at home safe in his stye, and not wandering about the hills. And what are these coming now?

Butter firkins, I think. Yes. This is a great flood. It is well if there are no lives lost.

But is it not cruel of Madam How to make such floods?

Well—let us ask one of these men who are looking over the bridge.

Why, what does he say? I cannot understand one word. Is he talking Irish?

Irish-English at least: but what he said was, that it was a mighty fine flood entirely, praised be God; and would help on the potatoes and oats after the drought, and set the grass growing again on the mountains.

And what is he saying now?

That the river will be full of salmon and white trout after this.

What does he mean?

That under our feet now, if we could see through the muddy water, dozens of salmon and sea-trout are running up from the sea.

What! up this furious stream?

Yes. What would be death to you is pleasure and play to them. Up they are going, to spawn in the little brooks among the mountains; and all of them are the best of food, fattened on the herrings and sprats in the sea outside, Madam How's free gift, which does not cost man a farthing, save the expense of nets and rods to catch them.

How can that be?

I will give you a bit of political economy. Suppose a pound of salmon is worth a shilling; and a pound of beef is worth a shilling likewise. Before we can eat the beef, it has cost perhaps tenpence to make that pound of beef out of turnips and grass and oil-cake; and so the country is only twopence a pound richer for it. But Mr. Salmon has made himself out of what he eats in the sea, and so has cost nothing; and the shilling a pound is all clear gain. There—you don't quite understand that piece of political economy. Indeed, it is only in the last two or three years that older heads than yours have got to understand it, and have passed the wise new salmon laws, by which the rivers will be once more as rich with food as the land is, just as they were hundreds of years ago. But now, look again at the river. What do you think makes it so yellow and muddy?

Dirt, of course.

And where does that come from?

Off the mountains?

Yes. Tons on tons of white mud are being carried down past us now; and where will they go?

Into the sea?

Yes, and sink there in the still water, to make new strata at the bottom; and perhaps in them, ages hence, some one will find the bones of those sheep, and of poor Mr. Pig too, fossil—

And the butter firkins too. What fun to find a fossil butter firkin!

But now lift up your eyes to the jagged mountain crests, and their dark sides all laced with silver streams. Out of every crack and cranny there aloft, the rain is bringing down dirt, and stones too, which have been split off by the winter's frosts, deepening every little hollow, and sharpening every peak, and making the hills more jagged and steep year by year.

When the ice went away, the hills were all scraped smooth and round by the glaciers, like the flat rock upon the lawn; and ugly enough they must have looked, most like great brown buns. But ever since then, Madam How has been scooping them out again by her water-chisel into deep glens, mighty cliffs, sharp peaks, such as you see aloft, and making the old hills beautiful once more. Why, even the Alps in Switzerland have been carved out by frost and rain, out of some great flat. The very peak of the Matterhorn, of which you have so often seen a picture, is but one single point left of some enormous bun of rock. All the rest has been carved away by rain and frost; and some day the Matterhorn itself will be carved away, and its last stone topple into the glacier at its foot. See, as we have been talking, we have got into the woods.

Oh, what beautiful woods, just like our own.

Not quite. There are some things growing here which do not grow at home, as you will soon see. And there are no rocks at home, either, as there are here.

How strange, to see trees growing out of rocks! How do their roots get into the stone?

There is plenty of rich mould in the cracks for them to feed on—

"Health to the oak of the mountains; he trusts to the might of the rock-clefts.

Deeply he mines, and in peace feeds on the wealth of the stone."

How many sorts of trees there are—oak, and birch and nuts, and mountain-ash, and holly and furze, and heather.

And if you went to some of the islands in the lake up in the glen, you would find wild arbutus—strawberry-tree, as you call it. We will go and get some one day or other.

How long and green the grass is, even on the rocks, and the ferns, and the moss, too. Everything seems richer here than at home.

FILM-FERN

Of course it is. You are here in the land of perpetual spring, where frost and snow seldom, or never comes.

Oh, look at the ferns under this rock! I must pick some.

Pick away. I will warrant you do not pick all the sorts.

Yes. I have got them all now.

Not so hasty, child; there is plenty of a beautiful fern growing among that moss, which you have passed over. Look here.

What! that little thing a fern!

Hold it up to the light, and see.

What a lovely little thing, like a transparent sea-weed, hung on black wire. What is it?

Film fern, Hymenophyllum. But what are you staring at now, with all your eyes?

Oh! that rock covered with green stars and a cloud of little white and pink flowers growing out of them.

Aha! my good little dog! I thought you would stand to that game when you found it.

What is it, though?

You must answer that yourself. You have seen it a hundred times before.

Why, it is London Pride, that grows in the garden at home.

Of course it is: but the Irish call it St. Patrick's cabbage; though it got here a long time before St. Patrick; and St. Patrick must have been very short of garden-stuff if he ever ate it.

But how did it get here from London?

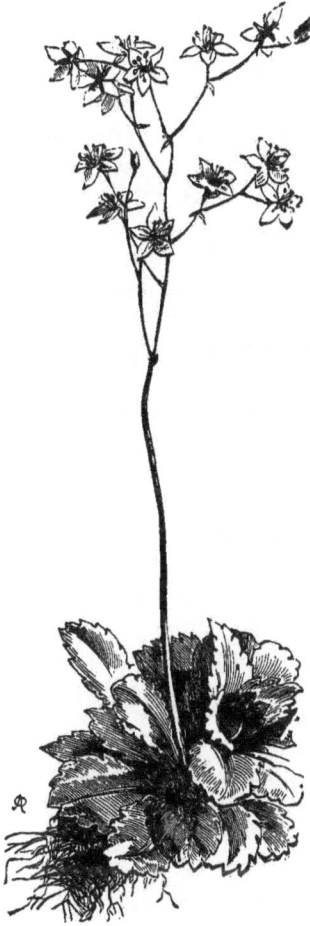

LONDON PRIDE

No, no. How did it get to London from hence? For from this country it came. I suppose the English brought it home in Queen Bess's or James the First's time.

But if it is wild here, and will grow so well in England, why do we not find it wild in England too?

For the same reason that there are no toads or snakes in Ireland. They had not got as far as Ireland before Ireland was parted off from England. And St. Patrick's cabbage, and a good many other plants, had not got as far as England.

But why?

Why, I don't know. But this I know: that when Madam How makes a new sort of plant or animal, she starts it in one single place, and leaves it to take care of itself and earn its own living—as she does you and me and every one—and spread from that place all round as far as it can go. So St. Patrick's cabbage got into this south-west of Ireland, long, long ago; and was such a brave sturdy little plant, that it clambered up to the top of the highest mountains, and over all the rocks. But when it got to the rich lowlands to the eastward, in county Cork, it found all the ground taken up already with other plants; and as they had enough to do to live themselves, they would not let St. Patrick's cabbage settle among them; and it had to be content with living here in the far-west—and, what was very sad, had no means of sending word to its brothers and sisters in the Pyrenees how it was getting on.

What do you mean? Are you making fun of me?

Not the least. I am only telling you a very strange story, which is literally true. Come, and sit down on this bench. You can't catch that great butterfly, he is too strong on the wing for you.

But oh, what a beautiful one!

Yes, orange and black, silver and green, a glorious creature. But you may see him at home sometimes: that plant close to you, you cannot see at home.

Why, it is only great spurge, such as grows in the woods at home.

No. It is Irish spurge which grows here, and sometimes in Devonshire, and then again in the west of Europe, down to the Pyrenees. Don't touch it. Our wood spurge is poisonous enough, but this is worse still; if you get a drop of its milk on your lip or eye, you will be in agonies for half a day. That is the evil plant with which the poachers kill the salmon.

How do they do that?

When the salmon are spawning up in the little brooks, and the water is low, they take that spurge, and grind it between two stones under water, and let the milk run down into the pool; and at that all the poor salmon turn up dead. Then comes the water-bailiff, and catches the poachers. Then comes the policeman, with his sword at his side and his truncheon under his arm: and then comes a "cheap journey" to Tralee Gaol, in which those foolish poachers sit and reconsider themselves, and determine not to break the salmon laws—at least till next time.

But why is it that this spurge, and St. Patrick's cabbage, grow only here in the west? If they got here of themselves, where did they come from? All outside there is sea; and they could not float over that.

Come, I say, and sit down on this bench, and I will tell you a tale,—the story of the Old Atlantis, the sunken land in the far West. Old Plato, the Greek, told legends of it, which you will read some day;

and now it seems as if those old legends had some truth in them, after all. We are standing now on one of the last remaining scraps of the old Atlantic land. Look down the bay. Do you see far away, under, the mountains, little islands, long and low?

Oh, yes.

Some of these are old slate, like the mountains; others are limestone; bits of the old coral-reef to the west of Ireland which became dry land.

I know. You told me about it.

Then that land, which is all eaten up by the waves now, once joined Ireland to Cornwall, and to Spain, and to the Azores, and I suspect to the Cape of Good Hope, and what is stranger, to Labrador, on the coast of North America.

Oh! How can you know that?

Listen, and I will give you your first lesson in what I call Biogeology.

What a long word!

If you can find a shorter one I shall be very much obliged to you, for I hate long words. But what it means is,—Telling how the land has changed in shape, by the plants and animals upon it. And if you ever read (as you will) Mr. Wallace's new book on the Indian Archipelago, you will see what wonderful discoveries men may make about such questions if they will but use their common sense. You know the common pink heather—ling, as we call it?

Of course.

Then that ling grows, not only here and in the north and west of Europe, but in the Azores too; and, what is more strange, in Labrador. Now, as ling can neither swim nor fly, does not common sense tell you that all those countries were probably joined together in old times?

Well: but it seems so strange.

So it is, my child; and so is everything. But, as the fool says in Shakespeare—

> "A long time ago the world began,
> With heigh ho, the wind and the rain."

And the wind and the rain have made strange work with the poor old world ever since. And that is about all that we, who are not very much wiser than Shakespeare's fool, can say about the matter. But again—the London Pride grows here, and so does another saxifrage very like it, which we call Saxifraga Geum. Now, when I saw those two plants growing in the Western Pyrenees, between France and Spain, and with them the beautiful blue butterwort, which grows in these Kerry bogs— we will go and find some—what could I say but that Spain and Ireland must have been joined once?

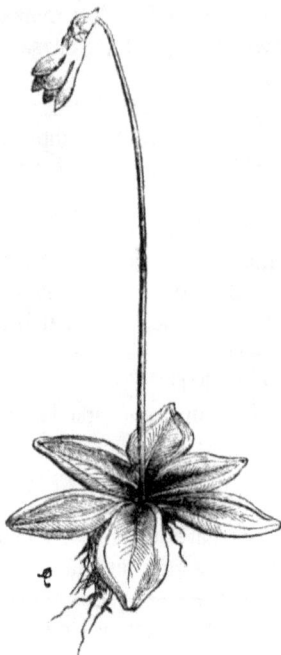

BUTTERWORT

I suppose it must be so.

Again. There is a little pink butterwort here in the bogs, which grows, too, in dear old Devonshire and Cornwall; and also in the south-west of Scotland. Now, when I found that too, in the bogs near Biarritz, close to the Pyrenees, and knew that it stretched away along the Spanish coast, and into Portugal, what could my common sense lead me to say but that Scotland, and Ireland, and Cornwall, and Spain were all joined once? Those are only a few examples. I could give you a dozen more. For instance, on an island away there to the west, and only in one spot, there grows a little sort of lily, which is found I believe in Brittany, and on the Spanish and Portuguese heaths, and even in Northwest Africa. And that Africa and Spain were joined not so very long ago at the Straits of Gibraltar there is no doubt at all.

But where did the Mediterranean Sea run out then?

Perhaps it did not run out at all; but was a salt-water lake, like the Caspian, or the Dead Sea. Perhaps it ran out over what is now the Sahara, the great desert of sand, for, that was a sea-bottom not long ago.

But then, how was this land of Atlantis joined to the Cape of Good

Hope?

I cannot say how, or when either. But this is plain: the place in the world where the most beautiful heaths grow is the Cape of Good Hope? You know I showed you Cape heaths once at the nursery gardener's at home.

Oh yes, pink, and yellow, and white; so much larger than ours.

Then it seems (I only say it seems) as if there must have been some land once to the westward, from which the different sorts of heath spread south-eastward to the Cape, and north-eastward into Europe. And that they came north-eastward into Europe seems certain; for there are no heaths in America or Asia.

IRISH HEATH

But how north-eastward?

Think. Stand with your face to the south and think. If a thing comes from the south-west—from there, it must go to the north-east-towards there. Must it not?

Oh yes, I see.

Now then—The farther you go south-west, towards Spain, the more kinds of heath there are, and the handsomer; as if their original home, from which they started, was somewhere down there.

More sorts! What sorts?

How many sorts of heath have we at home?

Three, of course: ling, and purple heath, and bottle heath.

And there are no more in all England, or Wales, or Scotland, except—Now, listen. In the very farthest end of Cornwall there are two more sorts, the Cornish heath and the Orange-bell; and they say (though I never saw it) that the Orange-bell grows near Bournemouth.

Well. That is south and west too.

So it is: but that makes five heaths. Now in the south and west of Ireland all these five heaths grow, and two more: the great Irish heath, with purple bells, and the Mediterranean heath, which flowers in spring.

Oh, I know them. They grow in the Rhododendron beds at home.

Of course. Now again. If you went down to Spain, you would find all those seven heaths, and other sorts with them, and those which are rare in England and Ireland are common there. About Biarritz, on the Spanish frontier, all the moors are covered with Cornish heath, and the bogs with Orange-bell, and lovely they are to see; and growing among them is a tall heath six feet high, which they call there *bruyère*, or Broomheath, because they make brooms of it: and out of its roots the "briar-root" pipes are made. There are other heaths about that country, too, whose names I do not know; so that when you are there, you fancy yourself in the very home of the heaths: but you are not. They must have come from some land near where the Azores are now; or how could heaths have got past Africa, and the tropics, to the Cape of Good Hope?

It seems very wonderful, to be able to find out that there was a great land once in the ocean all by a few little heaths.

Not by them only, child. There are many other plants, and animals too, which make one think that so it must have been. And now I will tell you something stranger still. There may have been a time—some people say that there must—when Africa and South America were joined by land.

Africa and South America! Was that before the heaths came here, or after?

I cannot tell: but I think, probably after. But this is certain, that there must have been a time when figs, and bamboos, and palms, and sarsaparillas, and many other sorts of plants could get from Africa to America, or the other way, and indeed almost round the world. About the south of France and Italy you will see one beautiful sarsaparilla, with hooked prickles, zigzagging and twining about over rocks and ruins, trunks and stems: and when you do, if you have understanding, it will seem as strange to you as it did to me to remember that the home of the sarsaparillas is not in Europe, but in the forests of Brazil, and the River Plate.

Oh, I have heard about their growing there, and staining the rivers brown, and making them good medicine to drink: but I never thought there were any in Europe.

There are only one or two, and how they got there is a marvel indeed. But now—If there was not dry land between Africa and South America, how did the cats get into America? For they cannot swim.

Cats? People might have brought them over.

Jaguars and Pumas, which you read of in Captain Mayne Reid's books, are cats, and so are the Ocelots or tiger cats.

Oh, I saw them at the Zoological Gardens.

But no one would bring them over, I should think, except to put them in the Zoo.

Not unless they were very foolish.

And much stronger and cleverer than the savages of South America. No, those jaguars and pumas have been in America for ages: and there are those who will tell you—and I think they have some reason on their side—that the jaguar, with his round patches of spots, was once very much the same as the African and Indian leopard, who can climb trees well. So when he got into the tropic forests of America, he took to the trees, and lived among the branches, feeding on sloths and monkeys, and never coming to the ground for weeks, till he grew fatter and stronger and far more terrible than his forefathers. And they will tell you, too, that the puma was, perhaps—I only say perhaps—something like the lion, who (you know) has no spots. But when he got into the forests, he found very little food under the trees, only a very few deer; and so he was starved, and dwindled down to the poor little sheep-stealing rogue he is now, of whom nobody is afraid.

Oh, yes! I remember now A. said he and his men killed six in one day. But do you think it is all true about the pumas and jaguars?

My child, I don't say that it is true: but only that it is likely to be true. In science we must be cautious and modest, and ready to alter our minds whenever we learn fresh facts; only keeping sure of one thing, that the truth, when we find it out, will be far more wonderful than any notions of ours. See! As we have been talking we have got nearly home: and luncheon must be ready.

* * * * *

Why are you opening your eyes at me like the dog when he wants to go out walking?

Because I want to go out. But I don't want to go out walking. I want to go in the yacht.

In the yacht? It does not belong to me.

Oh, that is only fun. I know everybody is going out in it to see such a beautiful island full of ferns, and have a picnic on the rocks; and I know you are going.

Then you know more than I do myself.

But I heard them say you were going.

Then they know more than I do myself.

But would you not like to go?

I might like to go very much indeed; but as I have been knocked about at sea a good deal, and perhaps more than I intend to be again, it is no novelty to me, and there might be other things which I liked still better: for instance, spending the afternoon with you.

Then am I not to go?

I think not. Don't pull such a long face: but be a man, and make up your mind to it, as the geese do to going barefoot.

But why may I not go?

Because I am not Madam How, but your Daddy.

What can that have to do with it?

If you asked Madam How, do you know what she would answer in a moment, as civilly and kindly as could be? She would say—Oh yes, go by all means, and please yourself, my pretty little man. My world is the Paradise which the Irishman talked of, in which "a man might do what was right in the sight of his own eyes, and what was wrong too, as he liked it."

Then Madam How would let me go in the yacht?

Of course she would, or jump overboard when you were in it; or put your finger in the fire, and your head afterwards; or eat Irish spurge, and die like the salmon; or anything else you liked. Nobody is so indulgent as Madam How: and she would be the dearest old lady in the world, but for one ugly trick that she has. She never tells any one what is coming, but leaves them to find it out for themselves. She lets them put their fingers in the fire, and never tells them that they will get burnt.

But that is very cruel and treacherous of her.

My boy, our business is not to call hard names, but to take things as we find them, as the Highlandman said when he ate the braxy mutton. Now shall I, because I am your Daddy, tell you what Madam How would not have told you? When you get on board the yacht, you will think it all very pleasant for an hour, as long as you are in the bay. But presently you will get a little bored, and run about the deck, and disturb people, and want to sit here, there, and everywhere, which I should not like. And when you get beyond that headland, you will find the great rollers coming in from the Atlantic, and the cutter tossing and heaving as you never felt before, under a burning sun. And then my merry little young gentleman will begin to feel a little sick; and then very sick, and more miserable than he ever felt in his life; and wish a thousand times over that he was safe at home, even doing sums in long division; and he will give a great deal of trouble to various kind ladies—which no one has a right to do, if he can help it.

Of course I do not wish to be sick: only it looks such beautiful weather.

And so it is: but don't fancy that last night's rain and wind can have passed without sending in such a swell as will frighten you, when you see the cutter climbing up one side of a wave, and running down the other; Madam How tells me that, though she will not tell you yet.

Then why do they go out?

Because they are accustomed to it. They have come hither all round from Cowes, past the Land's End, and past Cape Clear, and they are not afraid or sick either. But shall I tell you how you would end this evening?—at least so I suspect. Lying miserable in a stuffy cabin, on a sofa, and not quite sure whether you were dead or alive, till you were bundled into a boat about twelve o'clock at night, when you ought to be safe asleep, and come home cold, and wet, and stupid, and ill, and lie in bed all to-morrow.

But will they be wet and cold?

I cannot be sure; but from the look of the sky there to westward, I think some of them will be. So do you make up your mind to stay with me. But if it is fine and smooth to-morrow, perhaps we may row down the bay, and see plenty of wonderful things.

But why is it that Madam How will not tell people beforehand what will happen to them, as you have told me?

Now I will tell you a great secret, which, alas! every one has not found out yet. Madam How will teach you, but only by experience. Lady Why will teach you, but by something very different—by something which has been called—and I know no better names for it— grace and inspiration; by putting into your heart feelings which no man, not even your father and mother, can put there; by making you quick to love what is right, and hate what is wrong, simply because they are right and wrong, though you don't know why they are right and wrong; by making you teachable, modest, reverent, ready to believe those who are older and wiser than you when they tell you what you could never find out for yourself: and so you will be prudent, that is provident, foreseeing, and know what will happen if you do so-and-so; and therefore what is really best and wisest for you.

But why will she be kind enough to do that for me?

For the very same reason that I do it. For God's sake. Because God is your Father in heaven, as I am your father on earth, and He does not wish His little child to be left to the hard teaching of Nature and Law, but to be helped on by many, many unsought and undeserved favours, such as are rightly called "Means of Grace;" and above all by the Gospel and good news that you are God's child, and that God loves you, and has helped and taught you, and will help you and teach you, in a thousand ways of which you are not aware, if only you will be a wise child, and listen to Lady Why, when she cries from her Palace of Wisdom, and the feast which she has prepared, "Whoso is simple let him turn in hither;" and says to him who wants understanding— "Come, eat of my bread, and drink of the wine which I have mingled."

"Counsel is mine, and sound wisdom: I am understanding; I have strength. By me kings reign, and princes decree justice. By me princes rule, and nobles, even all the judges of the earth. I love them that love me; and those that seek me early shall find me. Riches and honour are with me; yea, durable riches and righteousness."

Yes, I will try and listen to Lady Why: but what will happen if I do not?

That will happen to you, my child—but God forbid it ever should happen—which happens to wicked kings and rulers, and all men, even the greatest and cleverest, if they do not choose to reign by Lady Why's laws, and decree justice according to her eternal ideas of what is just, but only do what seems pleasant and profitable to themselves. On them Lady Why turns round, and says—for she, too, can be awful, ay dreadful, when she needs—

"Because I have called, and ye refused; I have stretched out my hand, and no man regarded; but ye have set at nought all my counsel, and would have none of my reproof—" And then come words so terrible, that I will not speak them here in this happy place: but what they mean is this:—

That these foolish people are handed over—as you and I shall be if we do wrong willfully—to Madam How and her terrible school-house, which is called Nature and the Law, to be treated just as the plants and animals are treated, because they did not choose to behave like men and children of God. And there they learn, whether they like or not, what they might have learnt from Lady Why all along. They learn the great law, that as men sow so they will reap; as they make their bed so they will lie on it: and Madam How can teach that as no one else can in earth or heaven: only, unfortunately for her scholars, she is apt to hit so hard with her rod, which is called Experience, that they never get over it; and therefore most of those who will only be taught by Nature and Law are killed, poor creatures, before they have learnt their lesson; as many a savage tribe is destroyed, ay and great and mighty nations too—the old Roman Empire among them.

And the poor Jews, who were carried away captive to Babylon?

Yes; they would not listen to Lady Why, and so they were taken in hand by Madam How, and were seventy years in her terrible school-house, learning a lesson which, to do them justice, they never forgot again. But now we will talk of something pleasanter. We will go back to Lady Why, and listen to her voice. It sounds gentle and cheerful enough just now. Listen.

What? is she speaking to us now?

Hush! open your eyes and ears once more, for you are growing sleepy with my long sermon. Watch the sleepy shining water, and the sleepy green mountains. Listen to the sleepy lapping of the ripple, and the sleepy sighing of the woods, and let Lady Why talk to you through them in "songs without words," because they are deeper than all words, till you, too, fall asleep with your head upon my knee.

But what does she say?

She says—"Be still. The fullness of joy is peace." There, you are fast asleep; and perhaps that is the best thing for you; for sleep will (so I am informed, though I never saw it happen, nor any one else) put fresh gray matter into your brain; or save the wear and tear of the old gray matter; or something else—when they have settled what it is to do: and if so, you will wake up with a fresh fiddle-string to your little fiddle of a brain, on which you are playing new tunes all day long. So much the better: but when I believe that your brain is you, pretty boy, then I shall believe also that the fiddler is his fiddle.

Chapter XII. Homeward Bound

Come: I suppose you consider yourself quite a good sailor by now?

Oh, yes. I have never been ill yet, though it has been quite rough again and again.

What you call rough, little man. But as you are grown such a very good sailor, and also as the sea is all but smooth, I think we will have a sail in the yacht to-day, and that a tolerably long one.

Oh, how delightful! but I thought we were going home; and the things are all packed up.

And why should we not go homewards in the yacht, things and all?

What, all the way to England?

No, not so far as that; but these kind people, when they came into the harbour last night, offered to take us up the coast to a town, where we will sleep, and start comfortably home to-morrow morning. So now you will have a chance of seeing something of the great sea outside, and of seeing, perhaps, the whale himself.

I hope we shall see the whale. The men say he has been outside the harbour every day this week after the fish.

Very good. Now do you keep quiet, and out of the way, while we are getting ready to go on board; and take a last look at this pretty place, and all its dear kind people.

And the dear kind dogs too, and the cat and the kittens.

* * * * *

Now, come along, and bundle into the boat, if you have done bidding every one good-bye; and take care you don't slip down in the ice-groovings, as you did the other day. There, we are off at last.

Oh, look at them all on the rock watching us and waving their handkerchiefs; and Harper and Paddy too, and little Jimsy and Isy, with their fat bare feet, and their arms round the dogs' necks. I am so sorry to leave them all.

Not sorry to go home?

No, but—They have been so kind; and the dogs were so kind. I am sure they knew we were going, and were sorry too.

Perhaps they were. They knew we were going away, at all events. They know what bringing out boxes and luggage means well enough.

Sam knew, I am sure; but he did not care for us. He was only uneasy because he thought Harper was going, and he should lose his shooting; and as soon as he saw Harper was not getting into the boat, he sat down and scratched himself, quite happy. But do dogs think?

Of course they do, only they do not think in words, as we do.

But how can they think without words?

That is very difficult for you and me to imagine, because we always think in words. They must think in pictures, I suppose, by remembering things which have happened to them. You and I do that in our dreams. I suspect that savages, who have very few words to express their thoughts with, think in pictures, like their own dogs. But that is a long story. We must see about getting on board now, and under way.

* * * * *

Well, and what have you been doing?

Oh, I looked all over the yacht, at the ropes and curious things; and then I looked at the mountains, till I was tired; and then I heard you and some gentleman talking about the land sinking, and I listened. There was no harm in that?

None at all. But what did you hear him say?

That the land must be sinking here, because there were peat-bogs everywhere below high-water mark. Is that true?

Quite true; and that peat would never have been formed where the salt water could get at it, as it does now every tide.

But what was it he said about that cliff over there?

He said that cliff on our right, a hundred feet high, was plainly once joined on to that low island on our left.

What, that long bank of stones, with a house on it?

That is no house. That is a square lump of mud, the last remaining bit of earth which was once the moraine of a glacier. Every year it crumbles into the sea more and more; and in a few years it will be all gone, and nothing left but the great round boulder-stones which the ice brought down from the glaciers behind us.

But how does he know that it was once joined to the cliff?

Because that cliff, and the down behind it, where the cows are fed, is made up, like the island, of nothing but loose earth and stones; and that is why it is bright and green beside the gray rocks and brown heather of the moors at its foot. He knows that it must be an old glacier moraine; and he has reason to think that moraine once stretched right across the bay to the low island, and perhaps on to the other shore, and was eaten out by the sea as the land sank down.

But how does he know that the land sank?

Of that, he says, he is quite certain; and this is what he says.— Suppose there was a glacier here, where we are sailing now: it would end in an ice cliff, such as you have seen a picture of in Captain Cook's Voyages, of which you are so fond. You recollect the pictures of Christmas Sound and Possession Bay?

Oh yes, and pictures of Greenland and Spitzbergen too, with glaciers in the sea.

Then icebergs would break off from that cliff, and carry all the dirt and stones out to sea, perhaps hundreds of miles away, instead of letting it drop here in a heap; and what did fall in a heap here the sea would wash down at once, and smooth it over the sea-bottom, and never let it pile up in a huge bank like that. Do you understand?

I think I do.

Therefore, he says, that great moraine must have been built upon dry land, in the open air; and must have sunk since into the sea, which is gnawing at it day and night, and will some day eat it all up, as it would eat up all the dry land in the world, if Madam How was not continually lifting up fresh land, to make up for what the sea has carried off.

Oh, look there! some one has caught a fish, and is hauling it up. What a strange creature! It is not a mackerel, nor a gurnet, nor a pollock.

How do you know that?

Why, it is running along the top of the water like a snake; and they never do that. Here it comes. It has got a long beak, like a snipe. Oh, let me see.

See if you like: but don't get in the way. Remember you are but a little boy.

What is it? a snake with a bird's head?

No: a snake has no fins; and look at its beak: it is full of little teeth, which no bird has. But a very curious fellow he is, nevertheless: and his name is Gar-fish. Some call him Green-bone, because his bones are green.

But what kind of fish is he? He is like nothing I ever saw.

I believe he is nearest to a pike, though his backbone is different from a pike, and from all other known fishes.

But is he not very rare?

Oh no: he comes to Devonshire and Cornwall with the mackerel, as he has come here; and in calm weather he will swim on the top of the water, and play about, and catch flies, and stand bolt upright with his long nose in the air; and when the fisher-boys throw him a stick, he will jump over it again and again, and play with it in the most ridiculous way.

And what will they do with him?

Cut him up for bait, I suppose, for he is not very good to eat.

Certainly, he does smell very nasty.

Have you only just found out that? Sometimes when I have caught one, he has made the boat smell so that I was glad to throw him overboard, and so he saved his life by his nastiness. But they will catch plenty of mackerel now; for where he is they are; and where they are, perhaps the whale will be; for we are now well outside the harbour, and running across the open bay; and lucky for you that there are no rollers coming in from the Atlantic, and spouting up those cliffs in columns of white foam.

* * * * *

"Hoch!"

Ah! Who was that coughed just behind the ship?

Who, indeed? look round and see.

There is nobody. There could not be in the sea.

Look—there, a quarter of a mile away.

Oh! What is that turning over in the water, like a great black wheel? And a great tooth on it, and—oh! it is gone!

Never mind. It will soon show itself again.

But what was it?

The whale: one of them, at least; for the men say there are two different ones about the bay. That black wheel was part of his back, as he turned down; and the tooth on it was his back-fin.

But the noise, like a giant's cough?

Rather like the blast of a locomotive just starting. That was his breath.

What? as loud as that?

Why not? He is a very big fellow, and has big lungs.

How big is he?

I cannot say: perhaps thirty or forty feet long. We shall be able to see better soon. He will come up again, and very likely nearer us, where those birds are.

I don't want him to come any nearer.

You really need not be afraid. He is quite harmless.

But he might run against the yacht.

He might: and so might a hundred things happen which never do. But I never heard of one of these whales running against a vessel; so I suppose he has sense enough to know that the yacht is no concern of his, and to keep out of its way.

But why does he make that tremendous noise only once, and then go under water again?

You must remember that he is not a fish. A fish takes the water in through his mouth continually, and it runs over his gills, and out behind through his gill-covers. So the gills suck-up the air out of the water, and send it into the fish's blood, just as they do in the newt-larva.

Yes, I know.

But the whale breathes with lungs like you and me; and when he goes under water he has to hold his breath, as you and I have.

What a long time he can hold it.

Yes. He is a wonderful diver. Some whales, they say, will keep under for an hour. But while he is under, mind, the air in his lungs is getting foul, and full of carbonic acid, just as it would in your lungs, if you held your breath. So he is forced to come up at last: and then out of his blowers, which are on the top of his head, he blasts out all the foul breath, and with it the water which has got into his mouth, in a cloud of spray. Then he sucks in fresh air, as much as he wants, and dives again, as you saw him do just now.

And what does he do under water?

Look—and you will see. Look at those birds. We will sail up to them; for Mr. Whale will probably rise among them soon.

Oh, what a screaming and what a fighting! How many sorts there are! What are those beautiful little ones, like great white swallows, with crested heads and forked tails, who hover, and then dip down and pick up something?

Terns—sea-swallows. And there are gulls in hundreds, you see, large and small, gray-backed and black-backed; and over them all two or three great gannets swooping round and round.

Oh! one has fallen into the sea!

Yes, with a splash just like a cannon ball. And here he comes up again, with a fish in his beak. If he had fallen on your head, with that beak of his, he would have split it open. I have heard of men catching gannets by tying a fish on a board, and letting it float; and when the gannet strikes at it he drives his bill into the board, and cannot get it out.

But is not that cruel?

I think so. Gannets are of no use, for eating, or anything else.

What a noise! It is quite deafening. And what are those black birds about, who croak like crows, or parrots?

Look at them. Some have broad bills, with a white stripe on it, and cry something like the moor-hens at home. Those are razor-bills.

And what are those who say "marrock," something like a parrot?

The ones with thin bills? they are guillemots, "murres" as we call them in Devon: but in some places they call them "marrocks," from what they say.

And each has a little baby bird swimming behind it. Oh! there: the mother has cocked up her tail and dived, and the little one is swimming about looking for her! How it cries! It is afraid of the yacht.

And there she comes up again, and cries "marrock" to call it.

Look at it swimming up to her, and cuddling to her, quite happy.

Quite happy. And do you not think that any one who took a gun and shot either that mother or that child would be both cowardly and cruel?

But they might eat them.

These sea-birds are not good to eat. They taste too strong of fish-oil. They are of no use at all, except that the gulls' and terns' feathers are put into girls' hats.

Well they might find plenty of other things to put in their hats.

So I think. Yes: it would be very cruel, very cruel indeed, to do what some do, shoot at these poor things, and leave them floating about wounded till they die. But I suppose, if one gave them one's mind about such doings, and threatened to put the new Sea Fowl Act in force against them, and fine them, and show them up in the newspapers, they would say they meant no harm, and had never thought about its being cruel.

Then they ought to think.

They ought; and so ought you. Half the cruelty in the world, like half the misery, comes simply from people's not thinking; and boys are often very cruel from mere thoughtlessness. So when you are tempted to rob birds' nests, or to set the dogs on a moorhen, or pelt wrens in the hedge, think; and say—How should I like that to be done to me?

I know: but what are all the birds doing?

Look at the water, how it sparkles. It is alive with tiny fish, "fry," "brett" as we call them in the West, which the mackerel are driving up to the top.

Poor little things! How hard on them! The big fish at them from below, and the birds at them from above. And what is that? Thousands of fish leaping out of the water, scrambling over each other's backs. What a curious soft rushing roaring noise they make!

Aha! The eaters are going to be eaten in turn. Those are the mackerel themselves; and I suspect they see Mr. Whale, and are scrambling out of the way as fast as they can, lest he should swallow them down, a dozen at a time. Look out sharp for him now.

I hope he will not come very near.

No. The fish are going from us and past us. If he comes up, he will come up astern of us, so look back. There he is!

That? I thought it was a boat.

Yes. He does look very like a boat upside down. But that is only his head and shoulders. He will blow next.

"Hoch!"

Oh! What a jet of spray, like the Geysers! And the sun made a rainbow on the top of it. He is quite still now.

Yes; he is taking a long breath or two. You need not hold my hand so tight. His head is from us; and when he goes down he will go right away.

Oh, he is turning head over heels! There is his back fin again. And—Ah! was that not a slap! How the water boiled and foamed; and what a tail he had! And how the mackerel flew out of the water!

Yes. You are a lucky boy to have seen that. I have not seen one of those gentlemen show his "flukes," as they call them, since I was a boy on the Cornish coast.

Where is he gone?

Hunting mackerel, away out at sea. But did you notice something odd about his tail, as you call it—though it is really none?

It looked as if it was set on flat, and not upright, like a fish's. But why is it not a tail?

Just because it is set on flat, not upright: and learned men will tell you that those two flukes are the "rudiments"—that is, either the beginning, or more likely the last remains—of two hind feet. But that belongs to the second volume of Madam How's Book of Kind; and you have not yet learned any of the first volume, you know, except about a few butterflies. Look here! Here are more whales coming. Don't be frightened. They are only little ones, mackerel-hunting, like the big one.

What pretty smooth things, turning head over heels, and saying, "Hush, Hush!"

They don't really turn clean over; and that "Hush" is their way of breathing.

Are they the young ones of that great monster?

No; they are porpoises. That big one is, I believe, a bottle-nose. But if you want to know about the kinds of whales, you must ask Dr. Flower at the Royal College of Surgeons, and not me: and he will tell you wonderful things about them.—How some of them have mouths full of strong teeth, like these porpoises; and others, like the great sperm whale in the South Sea, have huge teeth in their lower jaws, and in the upper only holes into which those teeth fit; others like the bottle-nose, only two teeth or so in the lower jaw; and others, like the narwhal, two straight tusks in the upper jaw, only one of which grows, and is what you call a narwhal's horn.

SEA-BIRDS AND FISHES

Oh yes. I know of a walking-stick made of one.

And strangest of all, how the right-whales have a few little teeth when they are born, which never come through the gums; but, instead, they grow all along their gums, an enormous curtain of clotted hair, which serves as a net to keep in the tiny sea-animals on which they feed, and let the water strain out.

You mean whalebone? Is whalebone hair?

So it seems. And so is a rhinoceros's horn. A rhinoceros used to be hairy all over in old times: but now he carries all his hair on the end of his nose, except a few bristles on his tail. And the right-whale, not to be done in oddity, carries all his on his gums.

But have no whales any hair?

No real whales: but the Manati, which is very nearly a whale, has long bristly hair left. Don't you remember M.'s letter about the one he saw at Rio Janeiro?

This is all very funny: but what is the use of knowing so much about things' teeth and hair?

What is the use of learning Latin and Greek, and a dozen things more which you have to learn? You don't know yet: but wiser people than you tell you that they will be of use some day. And I can tell you, that if you would only study that gar-fish long enough, and compare him with another fish something like him, who has a long beak to his lower jaw, and none to his upper—and how he eats I cannot guess,— and both of them again with certain fishes like them, which M. Agassiz has found lately, not in the sea, but in the river Amazon; and then think carefully enough over their bones and teeth, and their history from the time they are hatched—why, you would find out, I believe, a story about the river Amazon itself, more wonderful than all the fairy tales you ever read.

Now there is luncheon ready. Come down below, and don't tumble down the companion-stairs; and by the time you have eaten your dinner we shall be very near the shore.

* * * * *

So? Here is my little man on deck, after a good night's rest. And he has not been the least sick, I hear.

Not a bit: but the cabin was so stuffy and hot, I asked leave to come on deck. What a huge steamer! But I do not like it as well as the yacht. It smells of oil and steam, and—

And pigs and bullocks too, I am sorry to say. Don't go forward above them, but stay here with me, and look round.

Where are we now? What are those high hills, far away to the left, above the lowlands and woods?

Those are the shore of the Old World—the Welsh mountains.

And in front of us I can see nothing but flat land. Where is that?

That is the mouth of the Severn and Avon; where we shall be in half an hour more.

And there, on the right, over the low hills, I can see higher ones, blue and hazy.

Those are an island of the Old World, called now the Mendip Hills; and we are steaming along the great strait between the Mendips and the Welsh mountains, which once was coral reef, and is now the

Severn sea; and by the time you have eaten your breakfast we shall
steam in through a crack in that coral-reef; and you will see what you
missed seeing when you went to Ireland, because you went on board at
night.

* * * * *

Oh! Where have we got to now? Where is the wide Severn Sea?

Two or three miles beyond us; and here we are in narrow little
Avon.

Narrow indeed. I wonder that the steamer does not run against
those rocks. But how beautiful they are, and how the trees hang down
over the water, and are all reflected in it!

Yes. The gorge of the Avon is always lovely. I saw it first when I
was a little boy like you; and I have seen it many a time since, in
sunshine and in storm, and thought it more lovely every time. Look!
there is something curious.

What? Those great rusty rings fixed into the rock?

Yes. Those may be as old, for aught I know, as Queen Elizabeth's
or James's reign.

But why were they put there?

For ships to hold on by, if they lost the tide.

What do you mean?

It is high tide now. That is why the water is almost up to the
branches of the trees. But when the tide turns, it will all rush out in a
torrent which would sweep ships out to sea again, if they had not steam,
as we have, to help them up against the stream. So sailing ships, in old
times, fastened themselves to those rings, and rode against the stream
till the tide turned, and carried them up to Bristol.

But what is the tide? And why does it go up and down? And why
does it alter with the moon, as I heard you all saying so often in
Ireland?

That is a long story, which I must tell you something about some
other time. Now I want you to look at something else: and that is, the
rocks themselves, in which the rings are. They are very curious in my
eyes, and very valuable; for they taught me a lesson in geology when I
was quite a boy: and I want them to teach it to you now.

What is there curious in them?

This. You will soon see for yourself, even from the steamer's deck,
that they are not the same rock as the high limestone hills above. They
are made up of red sand and pebbles; and they are a whole world
younger, indeed some say two worlds younger, than the limestone hills
above, and lie upon the top of the limestone. Now you may see what I
meant when I said that the newer rocks, though they lie on the top of
the older, were often lower down than they are.

But how do you know that they lie on the limestone?

Look into that corner of the river, as we turn round, and you will see with your own eyes. There are the sandstones, lying flat on the turned-up edges of another rock.

Yes; I see. The layers of it are almost upright.

Then that upright rock underneath is part of the great limestone hill above. So the hill must have been raised out of the sea, ages ago, and eaten back by the waves; and then the sand and pebbles made a beach at its foot, and hardened into stone; and there it is. And when you get through the limestone hills to Bristol, you will see more of these same red sandstone rocks, spread about at the foot of the limestone-hills, on the other side.

But why is the sandstone two worlds newer than the limestone?

Because between that sandstone and that limestone come hundreds of feet of rock, which carry in them all the coal in England. Don't you remember that I told you that once before?

Oh yes. But I see no coal between them there.

No. But there is plenty of coal between them over in Wales; and plenty too between them on the other side of Bristol. What you are looking at there is just the lip of a great coal-box, where the bottom and the lid join. The bottom is the mountain limestone; and the lid is the new red sandstone, or Trias, as they call it now: but the coal you cannot see. It is stowed inside the box, miles away from here. But now, look at the cliffs and the downs, which (they tell me) are just like the downs in the Holy Land; and the woods and villas, high over your head.

And what is that in the air? A bridge?

Yes—that is the famous Suspension Bridge—and a beautiful work of art it is. Ay, stare at it, and wonder at it, little man, of course.

But is it not wonderful?

Yes: it was a clever trick to get those chains across the gulf, high up in the air: but not so clever a trick as to make a single stone of which those piers are built, or a single flower or leaf in those woods. The more you see of Madam How's masonry and carpentry, the clumsier man's work will look to you. But now we must get ready to give up our tickets, and go ashore, and settle ourselves in the train; and then we shall have plenty to see as we run home; more curious, to my mind, than any suspension bridge.

And you promised to show me all the different rocks and soils as we went home, because it was so dark when we came from Reading.

Very good.

* * * * *

Now we are settled in the train. And what do you want to know first?

More about the new rocks being lower than the old ones, though they lie on the top of them.

Well, look here, at this sketch.

A boy piling up slates? What has that to do with it?

I saw you in Ireland piling slates against a rock just in this way. And I thought to myself—"That is something like Madam How's work."

How?

Why, see. The old rock stands for the mountains of the Old World, like the Welsh mountains, or the Mendip Hills. The slates stand for the new rocks, which have been piled up against these, one over the other. But, you see, each slate is lower than the one before it, and slopes more; till the last slate which you are putting on is the lowest of all, though it overlies all.

I see now. I see now.

Then look at the sketch of the rocks between this and home. It is only a rough sketch, of course: but it will make you understand something more about the matter. Now. You see, the lump marked A. With twisted lines in it. That stands for the Mendip Hills to the west, which are made of old red sandstone, very much the same rock (to speak roughly) as the Kerry mountains.

And why are the lines in it twisted?

To show that the strata, the layers in it, are twisted, and set up at quite different angles from the limestone.

But how was that done?

By old earthquakes and changes which happened in old worlds, ages on ages since. Then the edges of the old red sandstone were eaten away by the sea—and some think by ice too, in some earlier age of ice; and then the limestone coral reef was laid down on them, "unconformably," as geologists say—just as you saw the new red sandstone laid down on the edges of the limestone; and so one world is built up on the edge of another world, out of its scraps and ruins.

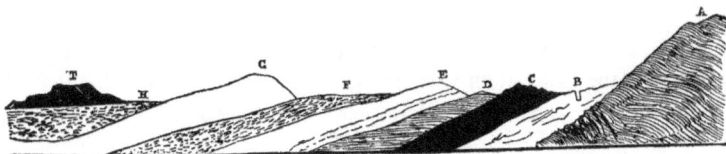

Then do you see B. With a notch in it? That means these limestone hills on the shoulder of the Mendips; and that notch is the gorge of the Avon which we have steamed through.

And what is that black above it?

That is the coal, a few miles off, marked C.

And what is this D, which comes next?

That is what we are on now. New red sandstone, lying unconformably on the coal. I showed it you in the bed of the river, as we came along in the cab. We are here in a sort of amphitheatre, or half a one, with the limestone hills around us, and the new red sandstone plastered on, as it were, round the bottom of it inside.

But what is this high bit with E against it?

Those are the high hills round Bath, which we shall run through soon. They are newer than the soil here; and they are (for an exception) higher too; for they are so much harder than the soil here, that the sea has not eaten them away, as it has all the lowlands from Bristol right into the Somersetshire flats.

* * * * *

There. We are off at last, and going to run home to Reading, through one of the loveliest lines (as I think) of old England. And between the intervals of eating fruit, we will geologize on the way home, with this little bit of paper to show us where we are.

What pretty rocks!

Yes. They are a boss of the coal measures, I believe, shoved up with the lias, the lias lying round them. But I warn you I may not be quite right: because I never looked at a geological map of this part of the line, and have learnt what I know, just as I want you to learn simply by looking out of the carriage window.

Look. Here is lias rock in the side of the cutting; layers of hard blue limestone, and then layers of blue mud between them, in which, if you could stop to look, you would find fossils in plenty; and along that lias we shall run to Bath, and then all the rocks will change.

* * * * *

Now, here we are at Bath; and here are the handsome fruit-women, waiting for you to buy.

And oh, what strawberries and cherries!

Yes. All this valley is very rich, and very sheltered too, and very warm; for the soft south-western air sweeps up it from the Bristol Channel; so the slopes are covered with fruit-orchards, as you will see as you get out of the station.

Why, we are above the tops of the houses.

Yes. We have been rising ever since we left Bristol; and you will soon see why. Now we have laid in as much fruit as is safe for you, and away we go.

Oh, what high hills over the town! And what beautiful stone houses! Even the cottages are built of stone.

All that stone comes out of those high hills, into which we are going now. It is called Bath-stone freestone, or oolite; and it lies on the top of the lias, which we have just left. Here it is marked F.

What steep hills, and cliffs too, and with quarries in them! What can have made them so steep? And what can have made this little narrow valley?

Madam How's rain-spade from above, I suppose, and perhaps the sea gnawing at their feet below. Those freestone hills once stretched high over our heads, and far away, I suppose, to the westward. Now they are all gnawed out into cliffs,—indeed gnawed clean through in the bottom of the valley, where the famous hot springs break out in which people bathe.

Is that why the place is called Bath?

Of course. But the Old Romans called the place Aquæ Solis—the waters of the sun; and curious old Roman remains are found here, which we have not time to stop and see.

Now look out at the pretty clear limestone stream running to meet us below, and the great limestone hills closing over us above. How do you think we shall get out from among them?

Shall we go over their tops?

No. That would be too steep a climb, for even such a great engine as this.

Then there is a crack which we can get through?

Look and see.

Why, we are coming to a regular wall of hill, and—

And going right through it in the dark. We are in the Box Tunnel.

* * * * *

There is the light again: and now I suppose you will find your tongue.

How long it seemed before we came out!

Yes, because you were waiting and watching, with nothing to look at: but the tunnel is only a mile and a quarter long after all, I believe. If you had been looking at fields and hedgerows all the while, you would have thought no time at all had passed.

What curious sandy rocks on each side of the cutting, in lines and layers.

Those are the freestone still: and full of fossils they are. But do you see that they dip away from us? Remember that. All the rocks are sloping eastward, the way we are going; and each new rock or soil we come to lies on the top of the one before it. Now we shall run down hill for many a mile, down the back of the oolites, past pretty Chippenham, and Wootton-Bassett, towards Swindon spire. Look at the country, child; and thank God for this fair English land, in which your lot is cast.

What beautiful green fields; and such huge elm trees; and orchards; and flowers in the cottage gardens!

Ay, and what crops, too: what wheat and beans, turnips and mangold. All this land is very rich and easily worked; and hereabouts is some of the best farming in England. The Agricultural College at Cirencester, of which you have so often heard, lies thereaway, a few miles to our left; and there lads go to learn to farm as no men in the world, save English and Scotch, know how to farm.

But what rock are we on now?

On rock that is much softer than that on the other side of the oolite hills: much softer, because it is much newer. We have got off the oolites on to what is called the Oxford clay; and then, I believe, on to the Coral rag, and on that again lies what we are coming to now. Do you see the red sand in that field?

Then that is the lowest layer of a fresh world, so to speak; a world still younger than the oolites—the chalk world.

But that is not chalk, or anything like it.

No, that is what is called Greensand.

But it is not green, it is red.

I know: but years ago it got the name from one green vein in it, in which the "Coprolites," as you learnt to call them at Cambridge, are found; and that, and a little layer of blue clay, called gault, between the upper Greensand and lower Greensand, runs along everywhere at the foot of the chalk hills.

I see the hills now. Are they chalk?

Yes, chalk they are: so we may begin to feel near home now. See how they range away to the south toward Devizes, and Westbury, and Warminster, a goodly land and large. At their feet, everywhere, run the rich pastures on which the Wiltshire cheese is made; and here and there, as at Westbury, there is good iron-ore in the greensand, which is being smelted now, as it used to be in the Weald of Surrey and Kent ages since. I must tell you about that some other time.

But are there Coprolites here?

I believe there are: I know there are some at Swindon; and I do not see why they should not be found, here and there, all the way along the foot of the downs, from here to Cambridge.

But do these downs go to Cambridge?

Of course they do. We are now in the great valley which runs right across England from south-west to north-east, from Axminster in Devonshire to Hunstanton in Norfolk, with the chalk always on your right hand, and the oolite hills on your left, till it ends by sinking into the sea, among the fens of Lincolnshire and Norfolk.

But what made that great valley?

I am not learned enough to tell. Only this I think we can say—that once on a time these chalk downs on our right reached high over our heads here, and far to the north; and that Madam How pared them away, whether by icebergs, or by sea-waves, or merely by rain, I cannot tell.

Well, those downs do look very like sea-cliffs.

So they do, very like an old shore-line. Be that as it may, after the chalk was eaten away, Madam How began digging into the soils below the chalk, on which we are now; and because they were mostly soft clays, she cut them out very easily, till she came down, or nearly down, to the harder freestone rocks which run along on our left hand, miles away; and so she scooped out this great vale, which we call here the Vale of White Horse; and further on, the Vale of Aylesbury; and then the Bedford Level; and then the dear ugly old Fens.

Is this the Vale of White Horse? Oh, I know about it; I have read *The Scouring of the White Horse.*

Of course you have; and when you are older you will read a jollier book still,—*Tom Brown's School Days*—and when we have passed Swindon, we shall see some of the very places described in it, close on our right.

* * * * *

There is the White Horse Hill.

The White Horse Hill? But where is the horse? I can see a bit of him: but he does not look like a horse from here, or indeed from any other place; he is a very old horse indeed, and a thousand years of wind and rain have spoilt his anatomy a good deal on the top of that wild down.

And is that really where Alfred fought the Danes?

As certainly, boy, I believe, as that Waterloo is where the Duke fought Napoleon. Yes: you may well stare at it with all your eyes, the noble down. It is one of the most sacred spots on English soil.

Ah, it is gone now. The train runs so fast.

So it does; too fast to let you look long at one thing: but in return, it lets you see so many more things in a given time than the slow old coaches and posters did.—Well? what is it?

I wanted to ask you a question, but you won't listen to me.

Won't I? I suppose I was dreaming with my eyes open. You see, I have been so often along this line—and through this country, too, long before the line was made—that I cannot pass it without its seeming full of memories—perhaps of ghosts.

Of real ghosts?

As real ghosts, I suspect, as any one on earth ever saw; faces and scenes which have printed themselves so deeply on one's brain, that when one passes the same place, long years after, they start up again, out of fields and roadsides, as if they were alive once more, and need sound sense to send them back again into their place as things which are past for ever, for good and ill. But what did you want to know?

Why, I am so tired of looking out of the window. It is all the same: fields and hedges, hedges and fields; and I want to talk.

Fields and hedges, hedges and fields? Peace and plenty, plenty and peace. However, it may seem dull, now that the grass is cut; but you would not have said so two months ago, when the fields were all golden-green with buttercups, and the whitethorn hedges like crested waves of snow. I should like to take a foreigner down the Vale of Berkshire in the end of May, and ask him what he thought of old England. But what shall we talk about?

I want to know about Coprolites, if they dig them here, as they do at Cambridge.

I don't think they do. But I suspect they will some day.

But why do people dig them?

Because they are rational men, and want manure for their fields.

But what are Coprolites?

Well, they were called Coprolites at first because some folk fancied they were the leavings of fossil animals, such as you may really find in the lias at Lynn in Dorsetshire. But they are not that; and all we can say is, that a long time ago, before the chalk began to be made, there was a shallow sea in England, the shore of which was so covered with dead animals, that the bone-earth (the phosphate of lime) out of them crusted itself round every bone, and shell, and dead sea-beast on the shore, and got covered up with fresh sand, and buried for ages as a mine of wealth.

But how many millions of dead creatures, there must have been! What killed them?

We do not know. No more do we know how it comes to pass that this thin band (often only a few inches thick) of dead creatures should stretch all the way from Dorsetshire to Norfolk, and, I believe, up through Lincolnshire. And what is stranger still, this same bone-earth bed crops out on the south side of the chalk at Farnham, and stretches along the foot of those downs, right into Kent, making the richest hop lands in England, through Surrey, and away to Tunbridge. So that it seems as if the bed lay under the chalk everywhere, if once we could get down to it.

But how does it make the hop lands so rich?

Because hops, like tobacco and vines, take more phosphorus out of the soil than any other plants which we grow in England; and it is the washings of this bone-earth bed which make the lower lands in Farnham so unusually rich, that in some of them—the garden, for instance, under the Bishop's castle—have grown hops without resting, I believe, for three hundred years.

But who found out all this about the Coprolites?

Ah—I will tell you; and show you how scientific men, whom ignorant people sometimes laugh at as dreamers, and mere pickers up of useless weeds and old stones, may do real service to their country and their countrymen, as I hope you will some day.

There was a clergyman named Henslow, now with God, honoured by all scientific men, a kind friend and teacher of mine, loved by every little child in his parish. His calling was botany: but he knew something of geology. And some of these Coprolites were brought him as curiosities, because they had fossils in them. But he (so the tale goes) had the wit to see that they were not, like other fossils, carbonate of lime, but phosphate of lime—bone earth. Whereon he told the neighbouring farmers that they had a mine of wealth opened to them, if they would but use them for manure. And after a while he was listened to. Then others began to find them in the Eastern counties; and then another man, as learned and wise as he was good and noble—John Paine of Farnham, also now with God—found them on his own estate, and made much use and much money of them: and now tens of thousands of pounds' worth of valuable manure are made out of them every year, in Cambridgeshire and Bedfordshire, by digging them out of land which was till lately only used for common farmers' crops.

But how do they turn Coprolites into manure? I used to see them in the railway trucks at Cambridge, and they were all like what I have at home—hard pebbles.

They grind them first in a mill. Then they mix them with sulphuric acid and water, and that melts them down, and parts them into two things. One is sulphate of lime (gypsum, as it is commonly called), and which will not dissolve in water, and is of little use. But the other is what is called superphosphate of lime, which will dissolve in water; so that the roots of the plants can suck it up: and that is one of the richest of manures.

Oh, I know: you put superphosphate on the grass last year.

Yes. But not that kind; a better one still. The superphosphate from the Coprolites is good; but the superphosphate from fresh bones is better still, and therefore dearer, because it has in it the fibrine of the bones, which is full of nitrogen, like gristle or meat; and all that has been washed out of the bone-earth bed ages and ages ago. But you must learn some chemistry to understand that.

I should like to be a scientific man, if one can find out such really useful things by science.

Child, there is no saying what you might find out, or of what use you may be to your fellow-men. A man working at science, however dull and dirty his work may seem at times, is like one of those "chiffoniers," as they call them in Paris—people who spend their lives in gathering rags and sifting refuse, but who may put their hands at any moment upon some precious jewel. And not only may you be able to help your neighbours to find out what will give them health and wealth: but you may, if you can only get them to listen to you, save them from many a foolish experiment, which ends in losing money just for want of science. I have heard of a man who, for want of science, was going to throw away great sums (I believe he, luckily for him, never could raise the money) in boring for coal in our Bagshot sands at home. The man thought that because there was coal under the heather moors in the North, there must needs be coal here likewise, when a geologist could have told him the contrary. There was another man at Hennequin's Lodge, near the Wellington College, who thought he would make the poor sands fertile by manuring them with whale oil, of all things in the world. So he not only lost all the cost of his whale oil, but made the land utterly barren, as it is unto this day; and all for want of science.

And I knew a manufacturer, too, who went to bore an Artesian well for water, and hired a regular well-borer to do it. But, meanwhile he was wise enough to ask a geologist of those parts how far he thought it was down to the water. The geologist made his calculations, and said:

"You will go through so many feet of Bagshot sand; and so many feet of London clay; and so many feet of the Thanet beds between them and the chalk: and then you will win water, at about 412 feet; but not, I think, till then."

The well-sinker laughed at that, and said, "He had no opinion of geologists, and such-like. He never found any clay in England but what he could get through in 150 feet."

So he began to bore—150 feet, 200, 300: and then he began to look rather silly; at last, at 405—only seven feet short of what the geologist had foretold—up came the water in a regular spout. But, lo and behold, not expecting to have to bore so deep, he had made his bore much too small; and the sand out of the Thanet beds "blew up" into the bore, and closed it. The poor manufacturer spent hundreds of pounds more in trying to get the sand out, but in vain; and he had at last to make a fresh and much larger well by the side of the old one, bewailing the day when he listened to the well-sinker and not to the geologist, and so threw away more than a thousand pounds. And there is an answer to what you asked on board the yacht—What use was there in learning little matters of natural history and science, which seemed of no use at all? And now, look out again. Do you see any change in the country?

What?

Why, there to the left.

There are high hills there now, as well as to the right. What are they?

Chalk hills too. The chalk is on both sides of us now. These are the Chilterns, all away to Ipsden and Nettlebed, and so on across Oxfordshire and Buckinghamshire, and into Hertfordshire; and on again to Royston and Cambridge, while below them lies the Vale of Aylesbury; you can just see the beginning of it on their left. A pleasant land are those hills, and wealthy; full of noble houses buried in the deep beech-woods, which once were a great forest, stretching in a ring round the north of London, full of deer and boar, and of wild bulls too, even as late as the twelfth century, according to the old legend of Thomas à Becket's father and the fair Saracen, which you have often heard.

I know. But how are you going to get through the chalk hills? Is there a tunnel as there is at Box and at Micheldever?

No. Something much prettier than a tunnel and something which took a great many years longer in making. We shall soon meet with a very remarkable and famous old gentleman, who is a great adept at digging, and at landscape gardening likewise; and he has dug out a path for himself through the chalk, which we shall take the liberty of using also. And his name, if you wish to know it, is Father Thames.

I see him. What a great river!

Yes. Here he comes, gleaming and winding down from Oxford, over the lowlands, past Wallingford; but where he is going to it is not so easy to see.

Ah, here is chalk in the cutting at last. And what a high bridge. And the river far under our feet. Why we are crossing him again!

Yes; he winds more sharply than a railroad can. But is not this prettier than a tunnel?

Oh, what hanging-woods, and churches; and such great houses, and pretty cottages and gardens—all in this narrow crack of a valley!

Ay. Old Father Thames is a good landscape gardener, as I said. There is Basildon—and Hurley—and Pangbourne, with its roaring lasher. Father Thames has had to work hard for many an age before he could cut this trench right through the chalk, and drain the water out of the flat vale behind us. But I suspect the sea helped him somewhat, or perhaps a great deal, just where we are now.

The sea?

Yes. The sea was once—and that not so very long ago—right up here, beyond Reading. This is the uppermost end of the great Thames valley, which must have been an estuary—a tide flat, like the mouth of the Severn, with the sea eating along at the foot of all the hills. And if the land sunk only some fifty feet,—which is a very little indeed, child, in this huge, ever-changing world,—then the tide would come up to Reading again, and the greater part of London and the county of Middlesex be drowned in salt water.

How dreadful that would be!

Dreadful indeed. God grant that it may never happen. More terrible changes of land and water have happened, and are happening still in the world: but none, I think, could happen which would destroy so much civilization and be such a loss to mankind, as that the Thames valley should become again what it was, geologically speaking, only the other day, when these gravel banks, over which we are running to Reading, were being washed out of the chalk cliffs up above at every tide, and rolled on a beach, as you have seen them rolling still at Ramsgate.

Now here we are at Reading. There is the carriage waiting, and away we are off home; and when we get home, and have seen everybody and everything, we will look over our section once more.

But remember, that when you ran through the chalk hills to Reading, you passed from the bottom of the chalk to the top of it, on to the Thames gravels, which lie there on the chalk, and on to the London clay, which lies on the chalk also, with the Thames gravels always over it. So that, you see, the newest layers, the London clay and the gravels, are lower in height than the limestone cliffs at Bristol, and much lower than the old mountain ranges of Devonshire and Wales, though in geological order they are far higher; and there are whole worlds of strata, rocks and clays, one on the other, between the Thames gravels and the Devonshire hills.

But how about our moors? They are newer still, you said, than the London clay, because they lie upon it: and yet they are much higher than we are here at Reading.

Very well said: so they are, two or three hundred feet higher. But our part of them was left behind, standing up in banks, while the valley of the Thames was being cut out by the sea. Once they spread all over where we stand now, and away behind us beyond Newbury in Berkshire, and away in front of us, all over where London now stands.

How can you tell that?

Because there are little caps—little patches—of them left on the tops of many hills to the north of London; just remnants which the sea, and the Thames, and the rain have not eaten down. Probably they once stretched right out to sea, sloping slowly under the waves, where the mouth of the Thames is now. You know the sand-cliffs at Bournemouth?

Of course.

Then those are of the same age as the Bagshot sands, and lie on the London clay, and slope down off the New Forest into the sea, which eats them up, as you know, year by year and day by day. And here were once perhaps cliffs just like them, where London Bridge now stands.

* * * * *

There, we are rumbling away home at last, over the dear old heather-moors. How far we have travelled—in our fancy at least—since we began to talk about all these things, upon the foggy November day, and first saw Madam How digging at the sand-banks with her water-spade. How many countries we have talked of; and what wonderful questions we have got answered, which all grew out of the first question, How were the heather-moors made? And yet we have not talked about a hundredth part of the things about which these very heather-moors ought to set us thinking. But so it is, child. Those who wish honestly to learn the laws of Madam How, which we call Nature, by looking honestly at what she does, which we call Fact, have only to begin by looking at the very smallest thing, pin's head or pebble, at their feet, and it may lead them—whither, they cannot tell. To answer any one question, you find you must answer another; and to answer that you must answer a third, and then a fourth; and so on for ever and ever.

For ever and ever?

Of course. If we thought and searched over the Universe—ay, I believe, only over this one little planet called earth—for millions on millions of years, we should not get to the end of our searching. The more we learnt, the more we should find there was left to learn. All things, we should find, are constituted according to a Divine and Wonderful Order, which links each thing to every other thing; so that we cannot fully comprehend any one thing without comprehending all things: and who can do that, save He who made all things? Therefore our true wisdom is never to fancy that we do comprehend: never to make systems and theories of the Universe (as they are called) as if we had stood by and looked on when time and space began to be; but to remember that those who say they understand, show, simply by so saying, that they understand nothing at all; that those who say they see, are sure to be blind; while those who confess that they are blind, are sure some day to see. All we can do is, to keep up the childlike heart, humble and teachable, though we grew as wise as Newton or as Humboldt; and to follow, as good Socrates bids us, Reason whithersoever it leads us, sure that it will never lead us wrong, unless we have darkened it by hasty and conceited fancies of our own, and so have become like those foolish men of old, of whom it was said that the very light within them was darkness. But if we love and reverence and trust Fact and Nature, which are the will, not merely of Madam How, or even of Lady Why, but of Almighty God Himself, then we shall be really loving, and reverencing, and trusting God; and we shall have our reward by discovering continually fresh wonders and fresh benefits to man; and find it as true of science, as it is of this life and of the life to come—that eye hath not seen, nor ear heard, nor hath it entered into the heart of man to conceive, what God has prepared for those who love Him.

THE END